Rodent Transplant Medicine

Weihua Gong
Editor

Rodent Transplant Medicine

Springer

Editor
Weihua Gong, MD, PhD
Department of Surgery
Zhejiang University
Hangzhou
China

ISBN 978-94-024-0242-1 ISBN 978-94-017-9472-5 (eBook)
DOI 10.1007/978-94-017-9472-5
Springer Dordrecht Heidelberg New York London

Printed on acid-free paper

Springer is part of Springer Science+Business Media (www.springer.com)

To my beloved wife, our son and parents

Preface

With the rapid development of microsurgery and molecular biology, rodent transplant models have become increasingly popular. Transplantation in rodents is the most useful in studying transplant immunobiology, in which more immunological tools can be utilized. However, the highly demanding techniques not only hamper the wide use of rodent transplant models but also affect the production of reliable, comparable, reproducible data. Many surgical modifications can de facto cause a significant variation of allograft survival. Systematic description of commonly used transplant models of solid organs (kidney, heart, liver) in rodents may improve surgical quality and save surgical time as a consequence, which will reduce and refine animal use. Adoption of uniform surgical procedures can produce comparable data efficiently enhancing the research translation from bench to nonhuman primate, even bedside. Furthermore, master of microsurgical techniques will benefit trainees in many surgical specialties and surgical research as well.

Indeed, the innovation of technologies can interplay with the progress of science. The refinement of microsurgical techniques has significantly promoted the development of transplant immunology because it has made the feasibility of rodent transplant models, in which a large amount of monoclonal antibodies, reagents, knockout/knockin animals, and other immunological tools can be utilized, but not for other laboratory species such as rabbits, dogs, swines, and nonhuman primates.

Currently, commercialized knockout rats have become available for research. Compared with the mouse, the rat is much more representative and indicative for human physiology, metabolism, and disease. Rat is advantageous in its larger size and physiological, cognitive, and breeding characteristics. Rat physiology is more suitable for studying human pharmacokinetics, toxicity, and immunology. Furthermore, more surgical procedures can be performed, and blood collection can be done simply and frequently in the larger animal.

As above-described, establishment of mouse and rat transplant models is of great value in application for both scientific research and improvement of microsurgical techniques. *Rodent Transplant Medicine* by Dr. Weihua Gong et al. is directed at shedding light on the successful establishment of rodent transplant models and understanding rodent transplant biology. The chapters written by German, American, and Chinese experts are accomplished by clear illustrations. It will help the readers better grasp the microsurgical techniques of rodent transplant models.

In addition, the challenge of translational medicine from animal models to human situations exists in practice, which requires study models closely mimicking clinical scenarios. Other influencing factors including donor factors (prolonged warm/cold ischemic time, hypertension, diabetes, and brain death) and recipient factors (coinfections, immunological memory) should be taken into account. These aspects might be further discussed for the future in the forward.

Berlin/Germany Hans-Dieter Volk

Acknowledgement

Herein, I would like to acknowledge my close friends and colleagues, Dr. Thomas Thornley (USA), Dr. Moritz Schmelzle (Germany), Dr. Karoline Edtinger (Germany), Dr. Christian Denecke (Austria), Dr. Anja Reutzel-Selke (Germany).

I also appreciate the following Principal Investigators for their great supports and helps during my oversea studies, Professor Dr. Petra Reinke, Professor Dr. Hans-Dieter Volk, Professor Dr. Birgit Sawitzki, Privatdozent Dr. Andreas Pascher, Professor Dr. Peter Neuhaus, Professor Dr. Laurence A. Turka, Professor Dr. Terry B. Strom, Professor Dr. Simon C. Robson, Professor Dr. Tianhui Zhu, Professor Dr. Ana J. Coito, Professor Dr. Jerzy Kupiec-Weglinski, Dr. Xiuda Shen, Professor Dr. Ronald W. Busuttil, Professor Dr. Stefan G. Tullius, Professor Dr. Xianchang Li, Professor Dr. Yuanli Liu, Professor Dr. Jing Ma.

Thanks to Beth Israel Deaconess Medical Center (Harvard Medical School), UCLA Medical Center, Rudolf Virchow Klinikum (Charité-University Berlin), and Chirurgische Klinik (University of Heidelberg) for offering me this opportunity to perform experimental studies and scientific research.

The authors acknowledge Professor Dr. Erdan Dong and Professor Dr. Ruijuan Sun for their kind encouragements. Thanks to Dr. Lida Su for preparing few illustrations and Mrs. James Hu for his great supports.

The work is partially supported by the Department of Education of Zhejiang Province (Grant No. Y201226017), the National Natural Science Foundation of China (No. 81270323, No. 81470527), the Qianjiang Talents Project of Zhejiang Province (Grant No.2012R10022), and Zhejiang Provincial Outstanding Youth Foundation (Grant No.LR13H020001).

Contents

Contributors

Jinzhen Cai, MD, PhD Department of Transplantation, Transplant Center, Tianjin First Central Hospital, Tianjin Medical University, Tianjin, People's Republic of China

Junwu Cai Department of Transplantation, Transplant Center, Tianjin First Central Hospital, Tianjin Medical University, Tianjin, People's Republic of China

Weihua Gong, MD, PhD Department of Surgery, The Second Affiliated Hospital of Zhejiang University School of Medicine, Hangzhou, People's Republic of China

I. Klein, MD Department of Surgery I, University of Wuerzburg, Wuerzburg, Germany

Xian Chang Li, MD, PhD Transplant Immunology Center, Methodist Hospital, Methodist Hospital Research Institute, Texas Medical Center, Houston, TX, USA

Dahai Liu, PhD Department of Biotechnology, Life Science College, Anhui University, Hefei, People's Republic of China

Guangxian Liu, MD, PhD Department of Urology and Kidney Transplant Center, Nanjing Drum Tower Hospital, The Affiliated Hospital of Nanjing University Medical School, Nanjing, Jiangsu, People's Republic of China

C. Otto, PhD Department of Surgery I, University of Wuerzburg, Wuerzburg, Germany

Ulrich Steger, MD Department of Surgery I, University of Wuerzburg, Würzburg, Germany

Qiquan Sun, MD, PhD Department of Renal Transplantation, The Third Affiliated Hospital of Sun Yat-sen University, Guangzhou, People's Republic of China

Abbreviations

AMW	Anterior median wall
AO	Aorta
CA	Common carotid artery
GVHD	Graft-versus-host disease
IHVC	Infrahepatic vena cava
IVC	Inferior vena cava
JV	External jugular vein
MHC	Major histocompatibility class
OLT	Orthotopic liver transplantation
PA	Pulmonary artery
SHVC	Suprahepatic vena cava
SVC	Superior vena cava

Chapter 1
Rodent Transplantation Medicine

Weihua Gong

1.1 Introduction

With a rapid advance of surgical instruments and modern healthcare quality, today clinical transplant surgeries are frequently and successfully performed over the past few decades. Nevertheless, microsurgical techniques are still required particularly for medical students and clinical surgical residencies. A microsurgery laboratory is suggested to provide a platform to set up a formal training program (Kizilisik et al. 1998), in which many clinical scenarios can be mimicked. More importantly, clinically relevant transplant models may be established in large measures to allow us to decipher underlying mechanisms of immunologic (innate, adaptive immunities, and signaling pathways) or nonimmunologic factors (effect of body weight, gender, ischemia-reperfusion, hypertension, and drug toxicity) although limitations of animal model exist in practice.

Indeed, it is well accepted that transplantation in rodents is the most useful in studying transplant immunobiology, in which more immunological tools can be utilized. However, the highly demanding techniques not only hamper the wide use of rodent transplant models but also affect the production of reliable, comparable, reproducible data. Many surgical modifications can give rise to a significant variation of allograft survival (Baldwin et al. 1973). Therefore, based upon published updated data, authors attempt to summarize some common transplant models in rodents in order to improve microsurgical quality and to save surgical time, as a consequence, which will reduce and refine animal use. Furthermore, adoption of uniform surgical procedures can produce comparable data, efficiently enhancing the research translation from bench to primate, even bedside.

W. Gong, MD, PhD
Department of Surgery, The Second Affiliated Hospital of Zhejiang University School of Medicine, Hangzhou, People's Republic of China
e-mail: weihua.gong@gmail.com

W. Gong (ed.), *Rodent Transplant Medicine*,
DOI 10.1007/978-94-017-9472-5_1

Indeed, the innovation of technologies can interplay with the progress of science. Since microsurgery was firstly introduced by Dr. Donaghy RMP et al. in the field of neurological surgery in 1962, more and more accurate, delicate, and small instruments including 10-0 and 11-0 suture strands have been developed (Jacobson et al. 1992). The refinement of microsurgical techniques has significantly promoted the development of transplant immunology because it has made the feasibility of rodent transplant models, in which a large amount of monoclonal antibodies, reagents, knockout/knockin animals, and other immunological tools can be utilized, but not for other laboratory species such as rabbits, dogs, swines, and primates. In practice, the first vascularized transplantation surgery for kidney and heart in dogs was performed by the French surgeon Alexis Carrel, who was awarded the 1912 Nobel Prize for his inaugural work. Dr. Joseph Murray did experimental transplant surgeries for dogs, showing the transplanted kidney could be induced to function. Later on, as a pioneer of clinical transplantation, Dr. Murray performed the first truly successful kidney transplantation from one identical twin (Ronald Herrick) to another (Richard Herrick) on December 23, 1954 at Peter Bent Brigham Hospital in Boston. His milestone work was then recognized with the Nobel Prize award for medicine in 1990.

As a pioneer of microsurgery, Dr. Sun Lee from the University of Pittsburgh was the first microsurgeon performing portacaval anastomosis in rat, marking an epoch in rodent transplant surgeries. Afterward, various vascularized transplantation techniques were developed in rodents including kidney, liver, heart, lung, and spleen.

In fact, the successful surgeries are based upon continuous practice and experience. The learning curve of consecutive transplant should be performed to achieve a stable success rate (Martins 2006). Every microsurgical procedure per se requires different learning curves and previous surgical experience may obviously affect the learning process. As the size of vessels of transplant organs becomes smaller, more surgeries are required for practice to reduce the incidence of mortality and postoperative complication such as bleeding and hypothermia. For instance, experience in rat transplant model will be instrumental in performing mice transplantation and shortening learning curve, however, which cannot guarantee success with stronger confidence. There are many potential pitfalls to be avoided during the entire transplant process. With respect to beginners, the first fifty operations are normally used for familiarizing the techniques (Martins 2006). Therefore, if transplant procedure can be standardized in a book, minimized operative time and fewer complications can be achieved.

1.2 Advantages of Microsurgical Training in the Laboratory

To minimize use of animals for experimental studies, it is of great importance for microsurgeons to shorten the learning curve, which will meet a moral and ethical need. In addition, microsurgical training in the laboratory will benefit resident surgeons from improvement of surgical techniques without time limitation, medicolegal responsibilities, and psychological pressures, which may stem from incidental mistakes (Pichierri et al. 2009). Experienced practice by using animals will prevent surgeons from stress, hurry, and anxiety, strengthening their confidence in precise movements of microsurgeries (Pichierri et al. 2009). Therefore, microsurgical training plays a critical role in either experimental or clinical surgeries for both young and senior microsurgeons. It is not skeptical that successful establishment of microsurgical laboratory in university hospitals can drive a development of surgical quality and scientific research (Martins and Montero 2006; Pichierri et al. 2009).

1.3 Availability of Knockout Rat for Transplant Research

Since microsurgical manipulations in mice are highly technical demanding, availability of knockout rat (a genetically engineered rat with a targeted mutation) will offer feasibility to easily establish various transplant models and understand human diseases. As a common laboratory animal, rat has been used for human disease research for over 150 years. In many circumstances, rats are superior to other laboratory animals for studying on human health problems. Compared with the mouse, rat is much more representative and indicative for human physiology, metabolism, and disease. Rat is advantageous in its larger size, physiological, cognitive, and breeding characteristics (Abbott 2004). Rat physiology is more suitable for studying human pharmacokinetics, toxicity, and immunology as the type and number of rat's detoxifying enzymes are similar to that in human (Lindblad-Toh 2004). For instance, rat's heart rate is more similar to that of human (Abbott 2004). Furthermore, more surgical procedures can be performed and blood collection can be done simply and frequently in the larger animal. In particular, commercialized knockout rats have already become available for research (Sigma-Aldrich, SAGE™ Labs). Availability of knockout rat is an alternative complementing existing transgenic mice. However, the weak point of utilization of rat models is an increase of experimental costs particularly for higher dose of biological agents such as expensive monoclonal antibodies.

1.4 Tolerability of Different Transplant Organs

During the past three decades, various therapeutic strategies have been developed for transplant tolerance induction. Each of them has been successfully validated in at least one rodent model although limitations of immunogenicity for CD4+/CD8+ subsets of T cells were shown in some rodent models (Lechler et al. 2005). Based on various success on rodent findings, studies would be extended to larger animals, nonhuman primates, and humans. Indeed, CD4+ or CD8+ subset alone is sufficient to reject allograft in the clinically relevant settings, which allows rodent models to be utilized for mimicking clinical scenarios. A variety of tissues or organs have been adopted, and a rough hierarchy of 'tolerability' exists as followings (from easiest to hardest): liver >> heart > kidney > islet >> skin (Lechler et al. 2005).

1.5 Vascularized Versus Nonvascularized Grafts

Vascularized allograft, such as heart, kidney, and liver, is characterized by continuous blood supply from recipient's circulation. Vascularized allograft is advantageous over nonvascularized allograft (skin) in studying on mechanism of graft rejection or transplant tolerance induction since nonvascularized graft is susceptible to nonspecific ischemic degeneration, resulting in tissue inflammation and necrosis and an aggravation of immune rejection. It even can be observed in nonvascularized isograft (Medawar 1944; Jones et al. 2001). However, the tempo of nonvascularized graft rejection (skin and islet) is faster than that of vascularized one (heart, kidney, and liver) (Lechler et al. 2005). One explanation is the presence of antigen-presenting cells (APCs) and tissue-specific antigens. In hematopoietic chimeras, the host did not tolerate skin graft from the same donor, but allogeneic hematopoietic cells (Steinmuller 2001). In an MHC ($H2^b$)-compatible and minor histocompatibility Ags-mismatched mouse model (BALB.B →C57BL/6), different immunoresponse was observed for skin and heart transplantation. A higher IFN-gamma response against H60 peptide and H60-specific CD8+T cells was found in heart transplantation, whereas a higher IFN-gamma response toward H4 peptide and H4-specific CD8+T cells occurred in skin recipient. This transplant model demonstrated the importance of primary vascularization for donor T-cell transmigration (Kwun et al. 2011). Indeed, the type and number of APCs within graft also play important role in rejection process. A considerable number of resident epidermal Langerhans cells (skin APCs) tune skin alloreactivity including efficient stimulation of alloreactive T cells although presence of Langerhans cells is not a prerequisite for skin rejection (Obhrai et al. 2008).

1.6 Comparison Between Mouse and Rat Transplant Models

Both rat and mouse are available to be used for experimental transplant studies. However, there are a couple of different issues to be clarified for utilization of rodents.

1. Requirement of microsurgical techniques is different between rat and mouse transplant models. Although great advances have been achieved in microsurgery and molecular biology, murine transplant models can be successfully established by a small number of research centers, implying that a higher level of technical demanding is required to establish transplants, such as small bowel, kidney, and liver in the mice. The diameter of mouse vascular lumen (0.4 mm) for anastomosis is 20-fold smaller compared with rat's (1–2 mm). Meticulous manipulations should be taken to prevent the vascular intima from impairing and subsequent thrombus formation (Ge and Gong 2011). To establish murine transplant models,

microsurgical skills in rat models are normally required, which will shorten learning curve and weaken feeling of frustration.

2. Rejection intensity of graft in the mice is less severe than that in the rat transplant model. Normally, spontaneous acceptance of kidney or liver can only occur in the rat low-responder strain combinations, whereas it can be frequently observed in all mouse strain combinations (Zhong 1999). This is a limitation of mouse model for transplant studies, in which hyperacute rejection is difficult to be produced even in the settings of presensitization. In mice, transplant tolerance may be induced using a few weeks of immunosuppression, whereas tolerance induction was hard despite administration of several decades of immunosuppression.
3. A greater variance of allograft survival can be observed in mice in comparison to rat transplant model. For instance, mean standard error of skin rat-to-mouse xenotransplantation using immunosuppression was 18.4 days, but it was only 2.3 days for mouse-to-rat skin xenografts (Zhong 1999).
4. Compared with human, rodents have a weaker complement system (Wehner et al. 2009). Although there is a similarity between rats and mice, mice' complement activity is remarkably weaker than rats'. For instance, in BALB/c mice serum complement activity is 50-fold lower than that in Lewis rats. This might partially explain the rare occurrence of hyperacute rejection in mouse transplant models and the generally weaker intensity of rejection in mice compared with rats. Some mouse strains are even deficient in complement. B1O.D2 old, DBA/2, AKR, A/J, these mice do not have C5. Although C4d and C3d are not absent in rats and mice, they are barely detected in heart allograft in immunoglobulin-deficient mice (Wehner et al. 2009). Aforementioned mice characteristics should be taken into account for experimental transplant studies.
5. Mice and rats have different responses to medications. Rats are more sensitive to cyclosporine (CsA) than mice. CsA at a dose of 100 ng/ml can induce tolerance of cardiac allograft in rats, whereas even ten-fold higher dose of CsA would not prevent heart allograft rejection in mice. Nevertheless, in terms of resistance of CsA, mouse model can mimic clinical scenario, in which a very high dose of CsA is only capable of partially suppressing calcineurin activity (Batiuk et al. 1996).
6. Different immune responses were observed in mice. Different patterns of cytokines can be produced by various strains of mice in response to certain antigen. For instance, BALB/c mice are poor in producing interferon-gamma, resulting in formation of a Th2-predominant immune response. This favors development of tolerance induction. Compared with Th2-prone BALB/c mice, fewer Th17 cells are generated by epicutaneous sensitization with protein antigen in Th1-prone C57BL/6 mice (Wang et al. 2009). In addition, induced strong Th2 response can be suppressed by low-energy visible light irradiation, but the weaker Th2 response in C57BL/6 can be enhanced (Chen et al. 2009). Furthermore, it is also importance in utilizing other different mouse strains for specific transplant biological studies (Tables 1.1 and 1.2).

Table 1.1 Spontaneous acceptance of mouse hearts, intestine, kidney, and liver allografts in different strain combinations (Zhong 1999)

Strain combinations	Heart	Intestine	Kidney	Liver
C57B6–BALB/C	0 % (0/14)	0 % (0/8)	20 % (2/10)	72 % (5/7)
C57B6–C3H	0 % (0/11)	0 % (0/6)	50 % (3/6)	73 % (8/11)
BALB/C–CBA	0 % (0/6)	0 % (0/5)	33 % (2/6)	57 % (4/7)

Table 1.2 Murine models of immunological deficiency

Mouse strains	Characteristics
SCID	Absence of functional T cells and B cells
RAG-1(–/–)	Absence of functional T cells and B cells
RAG-2(–/–)	Absence of functional T cells, B cells, and natural killer cells
Nude	Functional T-cell deficiency
129/Sv	Selective loss of all the B-cell function
Xid	B-cell abnormalities
Beige	Decreased activity of natural killer cells

1.7 Different Combinations of Rodents for Transplant Models

Inbred rodent strains and various organs are utilized for rodent transplant models based on different aims of studies. With respect to mice transplant models, congenic mouse strains are normally tested in different combinations involving strong H-2 histoincompatibilities. Congenic mice are genetically identical except major histocompatibility complex allele (class I and/or class II) such as C57BL/6 J (H-2^b), DBA/2 (H-2^d), BALB/c (H-2^d), C3H/He (H-2^k), CBA/Ca (H-2^k), CBA/J (H-2^k), and B10.A (H-2^a). The major histocompatibility complex (H-2 complex of mice, RT1 complex of rats)-mismatch between donor and recipient may elicit an allogeneic immune response, meeting a need for short-, medium-, long-term, or third-party survival studies (Ge and Gong 2011) (Tables 1.3, 1.4 and 1.5). For instance, engraft of BALB/c cardiac graft into C57BL/6 J recipient can induce acute rejection (Gong et al. 2012), whereas transplanting B6.C.H-2-bm12 (BM12) into a wild-type C57BL/6 causes chronic rejection of heart graft. In the latter model, only single MHC class II mismatch exists in the I-A^b locus of BM12 mice and survival time of cardiac graft exceeds 60 days with a chronic development of a significant vasculopathy (Bedi et al. 2010).

Table 1.3 Graft survival of different combinations of rat strains without any treatment in transplantation

Donor	Recipient	Graft	Graft survival (days)	Reference
MP rat	Wistar	Kidney	5×3, 6, 7×6, 8×2, 10×2	Sakagami (1976)
Brown-Norway	Lewis	Kidney	7, 7, 9, 13	Noris et al. (2001)
ACI rat	Lewis	Heart	6, 6, 6, 6, 7	Wang et al. (1994)
Brown-Norway	Lewis	Limb	4, 4, 5	Fealy et al. (1994)
DA (RT1a)	Lewis (RT1I)	Liver	11±2	Kunugi et al. (2011)
Lewis	Brown-Norway	Liver	7, 8, 10, 11, 11, 12, 14, 14	Qin et al. (2010)
ACI rat	Lewis	Liver	11.8±0.4	Redaelli et al. (2001)
PVG (RT1c)	Lewis (RT1l)	Liver	10, 11×2, 12, 13, 17, 19	Wang et al. (2010)
PVG (RT1c),	DA (RT1a)	Liver	>100×6	Wang et al. (2010)
WAG (RT1^{u})	BN (RT1^{n})	Small bowel	6×3, 7, 8, 8, 13, 14,14,15,15,19, 21, 24, 27, 31, 47	Saat et al. (1991)
Lewis (RT1^{l})	Fisher	Islet	4, 6, 5, 5, 5	Kim et al. (2008)

Table 1.4 Xenograft survival of different combinations of rodent/human strains without any treatment in transplantation

Donor	Recipient	Graft	Graft survival (days)	Reference
Wistar-Furth rat	C57BL/6 mouse	Islet	11, 11, 11, 14, 14, 27	Aebischer et al. (1991)
Human	Wistar	Islet	9, 9, 9, 8, 8, 8, 7, 7	Zhang et al. (2010)
Hamster	Lewis rat	Kidney	5, 5, 5, 5, 6, 6, 6, 6, 6, 6	Miyazawa et al. (1995)
Hamster	Lewis rat	Liver	3, 3, 3, 3, 3, 3	Murase et al. (1993)
Hamster	Sprague–Dawley	Heart	3, 3, 3, 4, 4	Zhang et al. (2007)

Table 1.5 Graft survival of different combinations of mouse strains without any treatment in transplantation

Donor	Recipient	Graft	Graft survival (days)	Reference
BALB/c	C3H/He	Heart	8, 9, 10, 10, 10, 11	Li et al. (1999)
C57BL/6	BALB/c	Heart	8×3, 9×6, 10, 11×3, 14	Zhang et al. (1996)
C57BL/6	C3H/HeN	Heart	10, 11, 13×3, 14×2, 15×2, 19, 20	Zhang et al. (1996)
BALB/c	CBA	Heart	10, 11, 13, 14, 15, 19	Zhang et al. (1996)
C57BL/6	BALB/c	Skin (tail)	7, 8, 8, 9, 9	Schwoebel et al. (2005)
BALB/c	C3H/He	Skin (tail)	8, 8, 9, 10, 10	Li et al. (1999)
C57BL/6	BALB/c	Liver	30, 32, >100×5	Zhang et al. (1996)
C57BL/6	C3H/HeN	Liver	30, 32, 50, >100×8	Zhang et al. (1996)
BALB/c	CBA	Liver	50×2, 63, >100×4	Zhang et al. (1996)
C57BL/6	BALB/c	Kidney	8×3, 9, 10, 28, 59, 63, >100×2	Zhang et al. (1996)
C57BL/6	C3H/HeN	Kidney	8, 9, 9, >100×3	Zhang et al. (1996)
BALB/c	CBA	Kidney	8, 9×2, >100×2	Zhang et al. (1996)
C57BL/6	BALB/c	Intestine	6×3, 7×3, 8, 19	Zhang et al. (1996)
C57BL/6	C3H/HeN	Intestine	6, 7×2, 8, 10, 11	Zhang et al. (1996)
BALB/c	CBA	Intestine	6, 7, 9, 10, 11	Zhang et al. (1996)

References

Abbott A. Laboratory animals: the Renaissance rat. Nature. 2004;428(6982):464–6.

Aebischer P, Lacy PE, Gerasimidi-Vazeou A, Hauptfeld V. Production of marked prolongation of islet xenograft survival (rat to mouse) by local release of mouse and rat antilymphocyte sera at transplant site. Diabetes. 1991;40(4):482–5.

Baldwin 3rd WM, Cohen N, Hrapchak BB. Prolonged survival of murine skin grafted across a weak histocompatibility barrier as a function of skin-grafting technique. Transplantation. 1973;15(4):419–22.

Batiuk TD, Urmson J, Vincent D, Yatscoff RW, Halloran PF. Quantitating immunosuppression. Estimating the 50% inhibitory concentration for in vivo cyclosporine in mice. Transplantation. 1996;61(11):1618–24.

Bedi DS, Riella LV, Tullius SG, Chandraker A. Animal models of chronic allograft injury: contributions and limitations to understanding the mechanism of long-term graft dysfunction. Transplantation. 2010;90(9):935–44.

Chen JS, Chiu HC, Hsu CJ, Liu CY, Hsieh PC, Miaw SC, Yu HS, Wang LF. Low-energy visible light irradiation modulates immune responses induced by epicutaneous sensitization with protein antigen. J Invest Dermatol. 2009;129(9):2258–64.

Fealy MJ, Umansky WS, Bickel KD, Nino JJ, Morris RE, Press BH. Efficacy of rapamycin and FK 506 in prolonging rat hind limb allograft survival. Ann Surg. 1994;219(1):88–93.

Ge F, Gong W. Strategies for successfully establishing a kidney transplant in a mouse model. Exp Clin Transplant. 2011;9(5):287–94.

Gong W, Thornley T, Whitcher GH, Ge F, Yuan S, Liu DJ, Balasubramanian S. Introduction of modified cervical cardiac transplant model in mice. Exp Clin Transplant. 2012;10(2):158–62.

Jacobson 2nd JH, Wallman LJ, Schumacher GA, Flanagan M, Suarez EL, Donaghy RM. Microsurgery as an aid to middle cerebral artery endarterectomy. Microsurgery. 1992;13(3): 112–7; discussion 117–8.

Jones ND, Turvey SE, Van Maurik A, Hara M, Kingsley CI, Smith CH, Mellor AL, Morris PJ, Wood KJ. Differential susceptibility of heart, skin, and islet allografts to T cell-mediated rejection. J Immunol. 2001;166(4):2824–30.

Kim YH, Lim DG, Wee YM, Kim JH, Yun CO, Choi MY, Park YH, Kim SC, Han DJ. Viral IL-10 gene transfer prolongs rat islet allograft survival. Cell Transplant. 2008;17(6):609–18.

Kizilisik A, al Sebayel M, Ramirez CB. Microsurgery training for transplantation research purposes. Transplant Proc. 1998;30(7):3016.

Kunugi S, Shimizu A, Ishii E, Kuwahara N, Arai T, Kataoka M, Masuda Y, Nagasaka S, Fukuda Y. The pathological characteristics of acute antibody-mediated rejection in DA-to-Lewis rat orthotopic liver transplantation. Transplant Proc. 2011;43(7):2737–40.

Kwun J, Malarkannan S, Burlingham WJ, Knechtle SJ. Primary vascularization of the graft determines the immunodominance of murine minor H antigens during organ transplantation. J Immunol. 2011;187(8):3997–4006.

Lechler RI, Sykes M, Thomson AW, Turka LA. Organ transplantation–how much of the promise has been realized? Nat Med. 2005;11(6):605–13.

Li Y, Li XC, Zheng XX, Wells AD, Turka LA, Strom TB. Blocking both signal 1 and signal 2 of T-cell activation prevents apoptosis of alloreactive T cells and induction of peripheral allograft tolerance. Nat Med. 1999;5(11):1298–302.

Lindblad-Toh K. Genome sequencing: three's company. Nature. 2004;428(6982):475–6.

Martins PN. Learning curve, surgical results and operative complications for kidney transplantation in mice. Microsurgery. 2006;26(8):590–3.

Martins PN, Montero EF. Organization of a microsurgery laboratory. Acta Cir Bras. 2006;21(3): 187–9.

Medawar PB. The behaviour and fate of skin autografts and skin homografts in rabbits: a report to the War Wounds Committee of the Medical Research Council. J Anat. 1944;78(Pt 5):176–99.

Miyazawa H, Murase N, Demetris AJ, Matsumoto K, Nakamura K, Ye Q, Manez R, Todo S, Starzl TE. Hamster to rat kidney xenotransplantation. Effects of FK 506, cyclophosphamide, organ perfusion, and complement inhibition. Transplantation. 1995;59(8):1183–8.

Murase N, Starzl TE, Demetris AJ, Valdivia L, Tanabe M, Cramer D, Makowka L. Hamster-to-rat heart and liver xenotransplantation with FK506 plus antiproliferative drugs. Transplantation. 1993;55(4):701–7; discussion 707–8.

Noris M, Cugini D, Casiraghi F, Azzollini N, De Deus Viera Moraes L, Mister M, Pezzotta A, Cavinato RA, Aiello S, Perico N, Remuzzi G. Thymic microchimerism correlates with the outcome of tolerance-inducing protocols for solid organ transplantation. J Am Soc Nephrol. 2001;12(12):2815–26.

Obhrai JS, Oberbarnscheidt M, Zhang N, Mueller DL, Shlomchik WD, Lakkis FG, Shlomchik MJ, Kaplan DH. Langerhans cells are not required for efficient skin graft rejection. J Invest Dermatol. 2008;128(8):1950–5.

Pichierri A, Frati A, Santoro A, Lenzi J, Delfini R, Pannarale L, Gaudio E, D'Andrea G, Cantore GP. How to set up a microsurgical laboratory on small animal models: organization, techniques, and impact on residency training. Neurosurg Rev. 2009;32(1):101–10; discussion 110.

Qin L, Guan HG, Zhou XJ, Yin J, Lan J, Qian HX. Blockade of 4-1BB/4-1BB ligand interactions prevents acute rejection in rat liver transplantation. Chin Med J (Engl). 2010;123(2): 212–5.

Redaelli CA, Wagner M, Tien YH, Mazzucchelli L, Stahel PF, Schilling MK, Dufour JF. 1 alpha,25-Dihydroxycholecalciferol reduces rejection and improves survival in rat liver allografts. Hepatology. 2001;34(5):926–34.

Saat RE, de Bruin RW, Heineman E, Jeekel J, Marquet RL. Total orthotopic allogeneic small bowel transplantation in rats: effect of allograft irradiation combined with cyclosporine-A therapy. Gut. 1991;32(6):654–6.

Sakagami K. Early prediction of acute rejection after inbred rat kidney transplantation using macrophage migration inhibition test. Acta Med Okayama. 1976;30(3):181–95.

Schwoebel F, Barsig J, Wendel A, Hamacher J. Quantitative assessment of mouse skin transplant rejection using digital photography. Lab Anim. 2005;39(2):209–14.

Steinmuller D. Skin allograft rejection by stable hematopoietic chimeras that accept organ allografts sill is an enigma. Transplantation. 2001;72(1):8–9.

Wang C, Tay SS, Tran GT, Hodgkinson SJ, Allen RD, Hall BM, McCaughan GW, Sharland AF, Bishop GA. Donor IL-4-treatment induces alternatively activated liver macrophages and IDO-expressing NK cells and promotes rat liver allograft acceptance. Transpl Immunol. 2010;22(3–4):172–8.

Wang J, Geissler EK, Fechner Jr JH, Burlingham WJ, Knechtle SJ. Use of donor serum to prevent passive transfer of hyperacute rejection. J Surg Res. 1994;57(1):150–5.

Wang LF, Chiu HC, Hsu CJ, Liu CY, Hsueh YH, Miaw SC. Epicutaneous sensitization with a protein antigen induces Th17 cells. J Dermatol Sci. 2009;54(3):192–7.

Wehner JR, Morrell CN, Rodriguez ER, Fairchild RL, Baldwin 3rd WM. Immunological challenges of cardiac transplantation: the need for better animal models to answer current clinical questions. J Clin Immunol. 2009;29(6):722–9.

Zhang J, Li H, Jiang N, Wang GY, Fu BS, Wang GS, Yang Y, Chen GH. Inhibition of rejection in murine islet xenografts by CTLA4Ig and CD40LIg gene transfer. Chin Med J (Engl). 2010;123(21):3106–9.

Zhang XG, Lu Y, Wang B, Li H, Yu L, Liu C, Wu Z, Liu XM. Cytokine production during the inhibition of acute vascular rejection in a concordant hamster-to-rat cardiac xenotransplantation model. Chin Med J (Engl). 2007;120(2):145–9.

Zhang Z, Zhu L, Quan D, Garcia B, Ozcay N, Duff J, Stiller C, Lazarovits A, Grant D, Zhong R. Pattern of liver, kidney, heart, and intestine allograft rejection in different mouse strain combinations. Transplantation. 1996;62(9):1267–72.

Zhong R. Organ transplantation in mice: current status and future prospects. Microsurgery. 1999;19(2):52–5.

Chapter 2
T-Cell Costimulatory Molecules

Qiquan Sun and Xian Chang Li

Transplant tolerance, similar to transplant rejection, is mediated by T cells, and activation of T cells is a prerequisite for both rejection and induction of donor-specific tolerance. Thus, costimulatory molecules play an important role in regulating rejection or acceptance of the allograft. Because of this, T-cell costimulatory molecules have been at the forefront of transplantation research for decades. The initial paradigm was that engagement of costimulatory molecules, in the context of TCR stimulation, promotes activation, clonal expansion, survival and effector differentiation of T cells, and therefore, their absence would have resulted in anergy and/or apoptotic cell death of activated T cells, a situation that favors transplant survival. Earlier studies, mostly using in vitro models, supported this paradigm. In fact, activation of T cells by TCR stimulation in the presence of CD28 costimulatory blockade often resulted in anergy and apoptosis of responding T cells. An implication of this finding is that by blocking costimulatory signals at a critical time of TCR stimulation, the T-cell response might be aborted, and in the setting of transplantation, rejection might be prevented, which may result in long-term allograft survival and possibly tolerance. Indeed, in selected rodent models, blocking CD28 and/or CD154 costimulation can induce prolonged allograft survival, and in some cases, donor-specific tolerance (Sayegh and Turka 1998). These initial findings generated tremendous enthusiasm toward clinical translation in that therapeutic blockade of key costimulatory pathways may be a promising way to the induction of transplant tolerance.

Q. Sun, MD, PhD
Department of Renal Transplantation, The Third Affiliated Hospital of Sun Yat-sen University, Guangzhou, People's Republic of China
e-mail: sunqiquan@hotmail.com

X.C. Li, MD, PhD (✉)
Transplant Immunology Center, Methodist Hospital, Methodist Hospital Research Institute, Texas Medical Center, Houston, TX, USA
e-mail: xcli@tmhs.org

W. Gong (ed.), *Rodent Transplant Medicine*,
DOI 10.1007/978-94-017-9472-5_2

As viewed from a high vintage point now, there are numerous successes and failures in targeting T-cell costimulatory pathways in transplant models, and studies in this area are also fraught with unexpected challenges. One of the first surprises came upon the discovery of CTLA-4, which is structurally related to CD28, but it inhibits rather than enhances T-cell activation. Importantly, CTLA-4 and CD28 bind to identical ligands on APCs (i.e., B7-1 and B7-2) (Sharpe and Freeman 2002). Thus, blockade of B7 might inhibit both stimulatory and inhibitory pathways. Since then, numerous additional costimulatory molecules have been discovered, which can either support or inhibit T-cell activation. Therefore, the original belief that costimulatory molecules are uniquely required for robust T-cell activation has undergone drastic revisions. We now know that it is the integration of both positive and negative costimulatory signals by T cells during and after their initial activation, dictated by their temporal and spatial expression patterns, that ultimately determines the fate and the function of activated T cells. Therefore, targeting costimulatory pathways for the purpose of allograft tolerance is more complex than simply blocking ligand-receptor interactions. Given these new findings, it is not surprising that blocking CD28/B7 or CD40/CD154 pathways has neither been universally effective in generating tolerance in rodent models nor has it been able to induce tolerance in nonhuman primates (Lechler et al. 2005).

The concept of T-cell costimulation has continued to evolve since its inception. It has been recognized that transplantation stimulates not only destructive T effector cells but also regulatory T cells and that it is the balance of T effector cells and Treg that ultimately determines the fate of an allograft (Lechler et al. 2003). It should be noted that not all T effector cells are functionally equivalent. Naïve CD4+ T cells can be further specialized into functionally distinct subsets (i.e., Th1, Th2, Th17, and Foxp3+ Tregs) depending on how they are (co)stimulated and the local cytokine milieu. Moreover, subsets of memory T cells can also be alloreactive, and both transition of T effector into memory T cells and re-activation of pre-existing memory T cells are also regulated, at least in part, by costimulatory molecules. Importantly, Foxp3+ Treg express many of the same costimulatory molecules that are also expressed on T effector/memory cells; however, engagement of identical costimulatory molecules on effector/memory T cells and Treg can have a strikingly different impact on their functions. Thus, it is critically important to resolve the exact functions of costimulatory molecules amongst different T-cell subsets during the evolution of the alloimmune response in vivo as new tolerogenic strategies are developed in the future.

Herein, we focus on costimulatory molecules in *the Ig superfamily*, *TNF/TNFR superfamily*, and *the TIM family*, reviewing advances in our understanding of their roles in regulating T effector cells and Treg. We will also address how different costimulatory molecules regulate subset specialization of T effector cells and how such responses might be modulated to promote transplant tolerance. Finally, we highlight new emerging approaches in tolerance induction by tipping the balance of Treg and T effector cells by selectively modulating costimulatory pathways that are critically involved.

2.1 The Ig Superfamily Costimulatory Molecules

Costimulatory molecules in the Ig superfamily are structurally related in that the extracellular region in these molecules contains similar Ig domains. Functionally, members in this family exhibit diverse and distinct effects on T-cell activation; they can either support (costimulatory) or inhibit (coinhibitory) T-cell activation and effector differentiation, and such opposing effects are related, at least in part, to the presence of an ITAM or an ITIM motif in their cytoplasmic region, because the ITAM and ITIM motifs have been shown to trigger stimulatory and inhibitory signals, respectively (Sharpe and Freeman 2002). In addition, some family members can also modify the homeostasis and function of Foxp3+ Treg, with which to exert effects on immune activation. Therefore, costimulatory molecules in this superfamily play an important role in both immune activation and active immune regulation; this has important therapeutic implications in tolerance induction in transplant settings.

2.1.1 CD28/CTLA-4/B7 Pathways

CD28 is a paramount costimulatory molecule in the Ig superfamily. It is a transmembrane protein that forms homodimers on the cell surface. In both mice and humans, CD28 is constitutively expressed on all naïve CD4+ and CD8+ T cells. CD28 has two known ligands, B7-1 (CD80) and B7-2 (CD86), both of which are type I transmembrane glycoproteins of the Ig superfamily. B7-1 expression is inducible on APCs and many other cell types including activated T cells and Foxp3+ Treg. In contrast, B7-2 is constitutively expressed by APCs, and its levels of expression can be upregulated under inflammatory conditions. In the presence of TCR stimulation, engagement of CD28 by B7 ligands triggers proliferation, survival, and cytokine production by activated T cells, therefore constituting an important pathway in driving effector T-cell responses. Accordingly, blockade of CD28 signaling in the presence of TCR engagement, best exemplified in studies with T-cell clones in vitro, induces T-cell anergy and apoptosis, which is the hallmark of T-cell costimulation as it was initially envisioned.

CTLA-4 is structurally related to CD28 and also binds to B7-1 and B7-2 ligands. However, CTLA-4 differs from CD28 in several key aspects. First, CTLA-4 is not expressed on resting T cells, and its expression is induced only upon T-cell activation. Foxp3+ Treg are an exception as they constitutively express CTLA-4 on the cell surface (Sakaguchi 2004). Also, CTLA-4 binds to B7 ligands with much higher affinity (10- to 20-fold) than CD28, with a preference for B7-1 molecule. Importantly, CTLA-4/B7 engagement potently suppresses T-cell activation and induces T-cell anergy, which is diametrically opposite to that of CD28/B7 engagement. In fact, CTLA-4$^{-/-}$ mice die prematurely from rampant lymphoproliferation and systemic inflammation. This striking phenotype is due to unopposed activation of CD4+ T cells in response to environmental antigens, and firmly establishes CTLA-4 as a central negative regulator in the immune system. In other models, antibody-mediated blockade of CTLA-4 negative signal markedly enhances immune responses, this approach has been actively exploited to enhance antitumor immunity (Bluestone 1997).

The role of CD28/CTLA-4/B7 pathways in transplant models has been extensively studied by many groups. It is well recognized that disruption of CD28/B7 pathway prolongs transplant survival in many, but not all, rodent models. However, targeting the CD28/B7 pathway alone rarely induces stable tolerance in rodents and is relatively ineffective in preventing rejection in large animal models. On the other hand, disruption of CTLA-4/B7 negative pathway antagonizes tolerance induction. Thus, blocking the CD28/B7 costimulation, and at the same time sparing or enhancing the CTLA-4/B7 coinhibitory signals, is absolutely important in tolerance induction. However, this has proven to be extremely challenging and it remains a key issue plaguing tolerance induction by blocking CD28/CTLA-4/B7 pathways. One reagent that has been widely used to target CD28 costimulation is CTLA-4Ig, a recombinant fusion protein consisting of the extracellular domain of CTLA-4 and the Fc portion of IgG. CTLA-4Ig takes advantage of the fact that CTLA-4 has a much higher binding affinity to B7 ligands than that of CD28, and therefore, acts as a competitive antagonist of CD28/B7 interactions. Initial studies demonstrated that CTLA-4Ig was remarkably effective in blocking T-cell proliferation and effector functions in response to alloantigen stimulation in vitro. In rodent transplant models, a brief course of CTLA-4Ig, especially when combined with donor-specific transfusion (DST), consistently prevented rejection of fully MHC-mismatched heart, kidney, and islet allografts, and in some cases, induced donor-specific tolerance. Similar results were also obtained using blocking mAbs against both B7-1 and B7-2 molecules, demonstrating the potency of targeting this pathway in prolonging transplant survival. However, when further tested in more stringent rodent models (e.g., skin transplantation) or in large animal models (e.g., nonhuman primates), CTLA-4Ig, with or without DST, completely failed to prevent transplant rejection (Kirk et al. 1997; Williams et al. 2000). The latter finding suggests that targeting CD28/B7 alone is insufficient to induce tolerance and that there must be other mechanisms involved in the rejection response when CD28/CTLA-4/B7 costimulation is targeted by CTLA-4Ig.

The failure of CD28/B7 blockade to consistently induce tolerance may also relate to the fact that CTLA-4Ig and anti-B7 mAbs also inhibit the ability of B7 ligands to interact with CTLA-4, a critical negative costimulatory molecule mentioned above. In several models, CTLA-4/B7 engagement is required for the induction of peripheral tolerance, and blocking CTLA-4/B7 interactions by any means can result in enhanced T-cell responses that prevent transplant tolerance (Markees et al. 1998). Approaches that selectively block B7-2 (B7-2 preferentially interacts with CD28) may in theory spare CTLA-4/B7-1 binding and enhance tolerance. However, CD28 can still engage B7-1 to deliver costimulatory signals to T effector cells. Another issue is that CD28 signaling plays a prominent role in the induction and homeostasis of Foxp3+ Treg, an issue that will be discussed separately below.

Thus, complete and prolonged CD28 blockade may not be desirable in the induction of Treg-mediated tolerance. As CTLA-4 signal is required for tolerance, approaches to enhance CTLA-4 signaling would be a useful therapeutic strategy. Unfortunately, harnessing the potency of this pathway has not been straightforward. CTLA-4 must be cross-linked by surface ligand to mediate negative signaling. To date, all soluble anti-CTLA-4 mAbs block, rather than enhance CTLA-4 signaling, thus resulting in augmented immune responsiveness. Using a different approach, it was shown that a tolerogenic anti-CD45RB mAb rapidly augments CTLA-4 expression on a subset of CD4 cells, suggesting a possibility that CTLA-4 can be specifically upregulated and might contribute to the tolerogenic effect of anti-CD45RB mAb. Indeed, blocking CTLA-4 using anti-CTLA-4 mAb precipitated acute rejection (Ariyan et al. 2003). A recent study provides more definitive evidence that CTLA-4 upregulation by anti-CD45RB is a prerequisite in tolerance induced by anti-CD45RB (Fecteau et al. 2001). Finally, CTLA4-Ig, which can interfere with CTLA-4/B7 interactions, inhibits engraftment mediated by anti-CD45RB in wild-type mice. These results demonstrate that upregulation of CTLA-4 expression and augmented CTLA-4 signaling can contribute to long-term engraftment and transplant tolerance. Another more direct approach to augment negative signaling through CTLA-4 has been to express a single-chain anti-CTLA-4 antibody on tumor cells, allowing CTLA-4 cross-linking to trigger negative signaling. When cotransferred with T cells into immunodeficient mice, the expression of anti-CTLA-4 by tumor cells inhibited tumor rejection. The utility of this approach in transplant models is uncertain, although it raises the possibility using gene therapy approach to trigger CTLA-4 negative signaling. Thus, targeting CD28/B7 pathway is more complex than initially anticipated. While CTLA-4Ig and anti-B7 mAbs have undeniable effect in extending graft survival in some models, they also have inherent limitations in tolerance induction because of interfering with CTLA-4 negative signaling.

In attempts to enhance efficacy, much effort has been focused on trying to pair CTLA-4Ig with blockade of other costimulatory pathways or with current immunosuppressive drugs to prevent transplant rejection. In general, CTLA4Ig can be successfully combined with blockade of other positive costimulatory pathways to enhance survival of fully MHC-mismatched transplants. Importantly, CTLA4-Ig might also be combined with certain immunosuppressive drugs commonly used in the clinic to prevent transplant rejection. This is a very attractive approach in designing clinical trials where one hopes to add experimental agents to existing regimens that would provide a margin of safety. An unexpected finding is that calcineurin inhibitors (cyclosporine and FK506) appear to antagonize the therapeutic effects of CTLA4-Ig, at least in fully MHC-mismatched transplant models. It should be noted that the antagonizing effect between calcineurin inhibitors and CTLA4-Ig is not obvious in minor antigen-mismatched transplants, where the alloreactive clone size is much smaller, suggesting that the number of responding T cells may influence the effect. One possible explanation is that calcineurin inhibitors interfere with activation-induced cell death of alloreactive T cells by costimulatory blockade, as apoptosis is required for peripheral tolerance induction, particularly when the alloreactive clone size is large (Li et al. 2001). Another explanation is that calcineurin inhibitors may block Treg homeostasis by inhibiting IL-2, thus interfering with tolerogenic strategies that depend on Treg induction. In another approach involving the induction of mixed chimerism where intrathymic deletion of donor reactive clones is the key mechanism of tolerance induction, calcineurin inhibitors do not interfere with the tolerogenic effects of CTLA4-Ig-based costimulatory blockade treatment (Kurtz et al. 2003). An important lesson from these studies is that not all immunosuppressive agents are additive or synergistic when combined with costimulatory blockade.

The observation that CTLA-4-Ig has not been able to produce long-term engraftment in preclinical models and that the efficacy of CTLA-4Ig can be antagonized by calcineurin inhibitors has led to an alternative and perhaps more pragmatic approach, which utilizes CTLA-4Ig not as a tolerogenic agent, but as an immunosuppressive agent in lieu of calcineurin inhibitors in the regimen. Currently, a mutant form of CTLA4-Ig, which is also termed LAE29Y or Belatacept, that has markedly prolonged association time with B7 molecules, is in Phase 3 clinical trials in human kidney transplantation. Two major trials include study of recipients receiving kidneys from extended criteria donors and studies enrolling recipients of kidney transplants from either living donors or deceased donors. Two additional studies with Belatacept are ongoing: a conversion study from calcineurin inhibitors to Belatacept and a rapid steroid withdrawal trial following induction with Thymoglobulin and maintenance therapy with Belatacept with either MMF or rapamycin (Vincenti and Kirk 2008). Finally, a trial supported by the Immune Tolerance Network will evaluate the effectiveness of a regimen consisting of Belatacept and Rapamycin to induce tolerance in recipients of kidney transplants from living donors (Vincenti and Kirk 2008). These trials are clearly a major step forward toward in bringing CD28/B7 costimulation blockade to the clinic.

Despite these advances, there remain significant challenges in targeting CD28/CTLA-4/B7 pathways in transplant models. In addition to delivering a negative signal in suppressing T-cell activation, CTLA-4 is also used by Foxp3+ Tregs as part of their suppressive repertoire. Both effects may benefit allograft survival. Given the opposite roles of CD28 and CTLA-4 in the T-cell response, strategies that can separately target CTLA-4 and CD28 are of great interests. As mentioned above, CD28/B7 blockade by CTLA-4Ig strongly inhibits T-cell responses and promotes anergy. However, interference with CTLA-4/B7 interactions by CTLA-4Ig could contribute to its lack of universal efficacy in rodent models. How the markedly prolonged binding of Belatacept (mutant CTLA-4Ig) for B7 affects CTLA-4 signaling and Treg homeostasis is currently under investigation. There are several other approaches to interfering with CD28 costimulation while preserving CTLA-4 signaling. One is to use anti-B7-2 to block the preferential binding of B7-2 with CD28, leaving B7-1 free to interact with its preferential ligand CTLA-4. This strategy has some success in murine models. However, CD28 can still interact with B7-1, so the upregulation of B7-1 and CTLA-4 must coincide in order for this approach to work, which is not always the case in vivo. Another approach is to try to develop blocking anti-CD28 mAbs. In rodent models, anti-CD28 mAb can promote activation induced cell death of T cells, and therefore are immunosuppressive. However, clinical trials using a superagonist anti-CD28 antibody as an immunosuppressive agent ended with disastrous consequences as several patients died of the side effects of massive T-cell activation in vivo (Suntharalingam et al. 2006). Moreover, the impact of anti-CD28 on homeostasis and function of Foxp3+ Treg is less well defined and clearly warrants further studies.

2.1.2 ICOS/B7h Pathway

ICOS (*i*nducible *co*stimulatory molecule) is a CD28 homolog, but unlike CD28, ICOS is not constitutively expressed on naïve T cells, and ICOS expression is induced upon T-cell activation. The ligand for ICOS is B7h (also named L-ICOS, B7RP-1, B7h-2, and GL-50). B7h is structurally related to B7 molecules, but it does not bind to CD28 and CTLA-4. B7h is expressed at low levels on resting APCs, B cells, and parenchymal cells including vascular endothelial cells, but its expression is rapidly upregulated upon cellular activation. Engagement of ICOS by B7h enhances T-cell proliferation, cytokine production, and survival. In some models, expression of ICOS is enhanced by CD28 costimulation, suggesting a potential crosstalk between CD28 and ICOS.

ICOS was initially thought to play an important role in regulating Th2 differentiation rather than the initial T-cell activation. For example, although an agonist anti-ICOS mAb enhanced proliferation of activated T cells, it markedly stimulated the expression Th2 cytokines such as IL-4, IL-5, and IL-10. This led to the belief that ICOS is an important costimulatory molecule for Th2 cells. Indeed, administration of an ICOS-Ig that blocks ICOS/B7h engagement suppressed Th2-mediated airway hyperreactivity. In other models, ICOS signaling is critically important in the development of IL-10-producing regulatory T cells that inhibit tissue inflammation. With the generation of ICOS knockout mice, the role of ICOS costimulation was much better defined. ICOS-deficient T cells are able to produce IFN-γ, but they fail to express IL-4 upon restimulation. Moreover, as compared to wild-type controls, a significantly higher number of ICOS-deficient CD4+ T cells retain the naïve phenotype ($CD44^{low}CD62L^{high}$) after activation (Dong et al. 2001; Tafuri et al. 2001). Finally, ICOS knockout mice exhibit a profound defect in Ig isotype class switching and germinal center formation, suggesting a key role for ICOS in B-cell function. Recently, it has been demonstrated that ICOS is critically involved in the differentiation of follicular T helper cells, which mediate B-cell activation and the formations of germinal centers.

In transplant models, ICOS blockade suppresses intragraft T-cell activation and cytokine expression, resulting in prolonged cardiac allograft survival in a manner similar to that seen in ICOS-deficient allograft recipients. ICOS blockade combined with anti-CD154 mAb prevented the development of chronic allograft vasculopathy. The exact mechanism for such a profound effect is not clear, but there are many variables that potentially affect the outcome of ICOS blockade in transplant settings. These variables include the timing of ICOS blockade following T-cell activation, the degree of MHC mismatch between the donor and recipient, the involvement of CD4+ versus of CD8+ T cells in rejection, as well as the involvement of CD28 costimulation. For example, in a heart transplant model in the mouse, blocking ICOS pathway early after transplantation had marginal effect on graft survival. However, delayed blockade of ICOS markedly extended the allograft survival. Similarly, ICOS knockout recipients showed prolonged heart allograft survival of similar magnitude to the early anti-ICOS mAb treatment group. Alloreactive CD4 T-cell proliferation and alloantibody production were suppressed in both treatment groups, but only delayed blockade resulted in suppression of effector CD8 T-cell generation. Similar results were obtained when delayed ICOS blockade was performed in $CD28^{-/-}$ recipients, indicating that ICOS is an important regulator of alloimmune responses even in the absence of CD28 signaling. Interestingly, an intact CTLA-4-signaling pathway and a normal STAT6 signaling are required for the therapeutic effects of ICOS blockade (Harada et al. 2003), suggesting that active regulation and a Th2 type of response may be critically important in this setting. The latter point is particularly interesting since in nontransplant models noted above, ICOS blockade often inhibits Th2 responses. Thus, there might be key differences concerning the role of ICOS in transplant models versus nontransplant models.

Taken together, the contribution of ICOS to T-cell-mediated immune responses and the consequences of ICOS blockade may be critically influenced by both the nature of the immune response and the timing of ICOS blockade. In addition, there are complex, yet important, interactions between the ICOS/B7h pathway and the CD28/CTLA-4/B7 pathways. It should be noted that ICOS is also expressed on memory T cells and Foxp3+ Treg. Blocking the ICOS/B7h pathway does not seem to affect memory T-cell activation or expansion, but reduces their recruitment to the allograft. The role of ICOS in Treg generation and function is largely unknown, but this is an important issue that requires further studies.

2.1.3 PD-1/PD-L1/PD-L2 Pathways

Programmed death-1 (PD-1) is a member of the Ig superfamily; it shares approximately 20 % homology with CTLA-4. PD-1 is expressed on activated CD4+ and CD8+ T cells as well as on activated B cells, NK cells, and macrophages. PD-1 interacts with two ligands, PD-L1 (B7-H1) and PD-L2 (B7-DC), both ligands are expressed on APCs, although with key differences in expression levels and kinetics. PD-L1 is constitutively expressed at low levels on APCs. PD-L1 is also expressed on a subset of activated T cells and Foxp3+ Tregs. Moreover, PD-L1 and PD-L2 can be expressed by a variety of parenchymal cells, including heart, lung, kidney, pancreas, and placenta, as well as on endothelial cells. In addition, the expression of PD-L1 and PD-L2 by parenchymal cells suggests that the parenchyma itself may actively participate in suppressing immune responses by engaging PD-1 negative signaling pathway. Earlier experiments demonstrate that engagement of PD-1 by its ligands inhibited T-cell proliferation and cytokine production in response to antigen stimulation, and the inhibitory effect was dependent on the strength of TCR stimulation, the presence or absence of CD28 costimulation or IL-2. The role of PD-1 as a negative regulator was later confirmed by the creation of PD-1 knockout mice. Genetic knockout of PD-1 results in an autoimmune phenotype with lupus-like glomerulonephritis and progressive arthritis. Interestingly, B cells and myeloid cells from PD-1-deficient mice are hyperresponsive to polyclonal stimulation in vitro, probably reflecting the broad expression of PD-1 in the immune system. It should be noted that compared to CTLA4 knockout mice, the autoimmune phenotype observed in PD-1-deficient mice is delayed, more limited, and its manifestations are strain dependent (Khoury and Sayegh 2004).

It has been shown that PD-1, PD-L1, and PD-L2 are prominently expressed in the rejecting heart allograft. In an in vivo assay, administration of blocking anti-PD-L1 mAb enhances alloantigen-driven T-cell expansion, promotes Th1 differentiation, and accelerates transplant rejection. In other studies, administration of a PD-L1-Ig fusion protein that triggers PD-1 negative signaling prevents allograft rejection and facilitates tolerance induction in the presence of CD154 blockade or suboptimal doses of rapamycin. Moreover, in combination with anti-CD154 therapy, PD-L1-Ig prevents the development of chronic allograft vasculopathy in cardiac transplants. In an acute GVHD model, PD-1 positive-infiltrating cells were found in increased frequency in multiple GVHD target organs (spleen, colon, and liver). Moreover, GVHD was exacerbated by administration of a blocking anti-PD-1 mAb. Collectively, these studies further confirm the negative regulatory functions of PD-1 in both organ transplants and graft-versus-host responses in vivo. Many studies suggest additional complexities concerning the PD-1/PD-L1/2 pathway in the regulation of alloimmunity. For example, in a murine heart transplant model, anti-PD-L1, but not anti-PD-1, accelerated heart allograft rejection in wild-type mice. Similarly, in a TCR transgenic model where allograft rejection is mediated by donor-specific CD4+ T cells, disruption of the PD-1/PD-L1, but not the PD-1/PD-L2 pathway, enhanced T-cell expansion and accelerated graft rejection. However, in CD28 knockout mice, both mAbs accelerated rejection. Thus, CD28 costimulation has considerable impact on the functionality of the PD-1 pathway. Moreover, blocking PD-L1 and PD-L2 can have strikingly different effects on in vivo activation of CD4+ and CD8+ T cells in response to alloantigen stimulation (Habicht et al. 2007). Again, this effect seems to be dependent on whether CD28 costimulation is present. Specifically, in a GVHD-like model where CFSE-labeled donor T cells are stimulated in vivo by host alloantigens, blocking PD-L2 but not PD-L1 enhanced proliferation of donor CD4+ and CD8+ T cells. In the absence of CD28 (donor T cells are from CD28 knockout mice), however, blocking PD-L2 accelerated CD8+ T-cell expansion but suppressed CD4+ T-cell proliferation in vivo (Habicht et al. 2007). Thus, in different settings, different PD-1 ligands are preferentially involved in regulating the immune response. This is likely due to differences in tissue expression of the ligands for PD-1. These studies demonstrate that this pathway may allow parenchymal cells to directly participate in regulation of the immune response in inflamed tissues.

Harnessing the potency of negative pathways to create tolerance remains an attractive approach as they represent the physiological means of damping or terminating immune responses. However, there are many issues that remain to be resolved concerning the role of PD-1/PD-L1/PD-L2 pathway in alloimmune responses. In some studies, PD-L1 and PD-L2 appear to trigger stimulatory rather than inhibitory signals to T cells and the basis for this difference is unclear. Memory T cells also express PD-1, and it remains possible that PD-1 may have different role in activation of naïve versus memory T cells. Also, the ligands for PD-1 have a wide tissue expression, and in some tissue they could have distinct roles. For example, transgenic expression of PD-L1 in pancreatic islets promotes spontaneous development of diabetes in B6 mice that are normally resistant to autoimmunity. Moreover, the PD-L1-expressing islet grafts are acutely rejected when transplanted into allogeneic hosts (Subudhi et al. 2004). In stark contrast, PD-L1 expression by parenchymal tissue of cardiac allografts inhibits rejection and prevents vasculopathy (Yang et al. 2008). These differences are so striking, and the basis for the differences is likely to be significant, and therefore warrants further experimentation. It has been recently shown that PD-L1 also binds B7-1 (Butte et al. 2007) and this new finding could help explain a negative signal previously thought to be delivered by B7 on T cells. Finally, Foxp3+ Tregs also express PD-L1. How PD-L1 regulates Treg function, homeostasis, and T-effector cells to Treg conversion deserves further investigation. Understanding these issues is clearly central to therapeutic targeting of PD-1 pathway in tolerance induction.

2.1.4 BTLA and CD160 Pathways

B and T lymphocyte attenuator (BTLA) belongs to the Ig superfamily; it is expressed by a variety of cell types including activated T cells, B cells, and APCs. Herpes virus entry mediator (HVEM) is the primary ligand for BTLA, which belongs to TNFR family. HVEM is reported to be expressed by activated T cells, B cells, as well as NK cells (Murphy et al. 2006). This is a typical example of a costimulatory pair formed between Ig and TNFR superfamily members. In the T-cell compartment, engagement of BTLA by HVEM can deliver an inhibitory signal, which proven to suppress T-cell activation by both in vitro and in vivo studies. However, the role of BTLA in regulating other cell types remains unknown. Beside BTLA, CD160, a glycosylphosphatidylinositol-anchored molecule of the Ig superfamily, is a newly identified ligand for HVEM. CD160 expressed on a subset of CD4+ T cells and CD8+ T cells. CD160/HVEM engagement can also deliver a negative signal, which inhibits T-cell activation. Notably, HVEM cross-linking with CD160 inhibits anti-CD3 and anti-CD28-induced T-cell proliferation, and inhibition of CD4+ T-cell proliferation by HVEM-transfected cells appears to be dependent on both CD160 and BTLA (Cai et al. 2008). HVEM also has another two ligands called LIGHT and LT-alpha (lymphotoxin-alpha). Unlike BTLA and CD160, the binding of LIGHT or LT-alpha to HVEM triggers a costimulatory signal. Thus, HVEM can trigger either a positive or negative signaling events depending on binding with different ligands.

The exact role of BTLA in transplant models remains incompletely defined. In heart transplant model, the effect of BTLA on transplant outcome may depend on the degree of MHC mismatch or interaction with other negative pathways, such as PD-1. Recipient BTLA deficiency, or BTLA blockade using a blocking mAb may induce prompt heart allograft rejection in a partial MHC class II mismatched heart transplant model (bm12 to B6), whereas wild-type recipients accepted the graft long term. However, the role of BTLA signaling in fully MHC-mismatched transplant remains to be determined. Thus, the exact role of BTLA and its costimulatory molecules in transplant models need to be further studied.

2.1.4.1 The TNF-TNFR Superfamily Costimulatory Molecules

This family consists of numerous receptor ligand pairs, and CD40 and its ligand, CD154, are the first pair of costimulatory molecules identified in the TNFR and TNF superfamily. Since then, this family has expanded dramatically. With few exceptions, members in this superfamily are expressed predominantly by activated T cells. This has led to the general view that costimulatory molecules in this superfamily may play a particularly important role in the effector and memory phases of an immune response rather than the initial phase of T-cell priming. Importantly, memory T cells and Foxp3+ Treg have been shown to constitutively express certain family members. These findings highlight the importance of these family members in regulating diverse T-cell subsets in the immune system.

2.1.5 CD40/CD154 Pathway

CD40 and CD154 are members in the TNFR and TNF families, respectively. CD40, member in the TNFR family, is constitutively expressed on APCs at low levels. In response to stimuli, its expression could be significantly upregulated. CD40 can be induced on parenchymal cells as well, such as endothelial cells and fibroblasts in inflamed tissues. CD154, member of TNF family, is the ligand for CD40. It is expressed on activated T cells, NK cells, and eosinophils. It is also expressed on platelets in some cases. CD40/CD154 engagement can promote a robust immune response via several mechanisms, one of which is through CD40-mediated activation of dendritic cells. Activated dendritic cells then in turn upregulate other molecules, such as B7, MHC class II, and produce potent proinflammatory cytokines. These cytokines will promote antigen presentation and productive immune responses (Yang and Wilson 1996).

In several rodent transplant models, an anti-CD154 mAb (MR1) has been used to blocking CD40/CD154 costimulation, which has proven to be remarkably effective in preventing acute allograft rejection. A brief course of anti-CD154 mAb can prolong the survival of heart, kidney, and islet allografts. A prolonged period of graft surviving time can also achieved in heart and islet transplant models with

CD154 knockout mice. However, although CD154 blockade (or genetic knockout) effectively preventing acute allograft rejection, it fails to prevent the development of chronic vasculopathy in cardiac transplant models alone. Thus, many methods have been used to augment the efficacy of CD154 blockade such as by combination with donor-specific transfusion (DST), CTLA4-Ig, and rapamycin. In islet and cardiac transplant models, anti-CD154 combining with DST routinely leads to donor-specific tolerance; nevertheless, chronic vasculopathy can be prevented in cardiac transplant models. Although CTLA-4-Ig synergizes with anti-CD154, CTLA-4-Ig synergizes with anti-CD154 is not effective at prolonging skin graft survival in stringent strain combinations. However, CD40/CD154 blockade plus DST works in this model; moreover, long-term engraftment could be achieved with addition of anti-CD45RB. Obviously, CD40/CD154 blockade is an important component of a commonly used nonmyeloablative conditioning regimen for the induction of mixed chimerism and donor-specific transplant tolerance.

Based on impressive rodent studies, many investigators are trying to translate the CD40/CD154 blockade strategy into clinical practice. Many studies had been performed in nonhuman primate models with exciting results. For example, treatment of rhesus monkeys with an intense anti-CD154 regimen could prolong graft survival in kidney and pancreatic islet transplantation. The graft could even survive for over a year after therapy was terminated in some cases. It is clear that CD40/CD154 blockade promote transplant survival. However, rejection will ultimately occur after withdrawing from the therapy, suggesting that CD40/CD154 blockade only induce robust immunosuppression, but not classical tolerance. Moreover, the biggest surprise of the CD40/CD154 blockade strategy was the high incidence of thromboembolic events in both primate studies and Phase 1 clinical trials. This unexpected complication is likely related to the expression of CD154 on platelets, which promotes platelet aggregation when bound by anti-CD154 mAb (Xu et al. 2006). Therefore, future development of anti-CD154 mAbs in the clinic has been halted because of safety concerns. Nonetheless, the remarkable effect of targeting CD40/CD154 pathway on transplant survival prompted the search of alternative options to circumvent problems associated with anti-CD154. There are tremendous efforts now in developing anti-CD40 mAbs in attempts to block the same pathway without triggering DC activation and platelet aggregation. Initial results suggest that available anti-CD40 mAbs are not as potent as anti-CD154, but encouraging results have been achieved in combination with other agents in graft survival.

It should be noted that alloreactive memory T cells, especially memory CD8+ T cells, do not use CD154 for activation or effector function. Thus, mechanisms related to graft damage in the absence of CD154 costimulation are likely mediated by the activation of memory T cells. Furthermore, blocking CD40/CD154 pathway does not directly affect Foxp3+ Treg. Also, CD154-deficient mice have comparable number of Foxp3+ Treg as wild-type control mice. Interestingly, blocking CD154 on T effector cells seems to render them more sensitive to Treg-mediated suppression. Whether this is a general feature or only confined to specific experimental conditions remains to be defined.

2.1.6 *OX40/OX40L Pathway*

OX40 (CD134) was initially identified as a T-cell activation marker, since it is abundantly expressed on activated but not resting T cells. The ligand for OX40 (OX40L) is expressed on dendritic cells, B cells, and inflamed vascular endothelial cells. OX40/OX40L engagement has a broad impact on the T-cell response. It delivers an important costimulatory signal to activated T cells, and in some models OX40 costimulation is as potent as CD28 in mediating T-cell proliferation and survival. OX40 costimulation is also involved in regulating the effector programs of activated T cells. Depending on the models studied, OX40 is critically important in the generation of either a Th1 or a Th2 program. Perhaps the most significant role of OX40 is in supporting the generation of memory T cells, especially memory CD4+ T cells. Studies using OX40 knockout and OX40L knockout mice have consistently demonstrated that absence of the OX40 signaling has minimal impact on the initial T-cell activation, but survival of activated T cells cannot be sustained and generation of memory T cells and memory recall responses to specific antigens are severely impaired. On the other hand, deliberately stimulating OX40 costimulation either using an agonist anti-OX40 mAb, or expressing OX40L as a transgene, can markedly expand the CD4+ memory pool in vivo. Finally, in some models, OX40 signaling allows anergic T cells to reacquire effector function. Thus, disruption of OX40/OX40L pathway is highly beneficial in several models of autoimmune diseases. Conversely, deliberately triggering the OX40 receptor significantly enhances protective immunity and antitumor immunity (Sugamura et al. 2004).

In transplant settings, blocking OX40/OX40L pathway using a blocking anti-OX40L mAb exhibits remarkable synergy with CD28 and/or CD154 blockade in preventing rejection of MHC-mismatched transplants. Initial studies in a rat heart transplant model demonstrated that treatment of recipients with anti-OX40L and CTLA-4Ig consistently produced long-term heart allograft survival. In a stringent skin transplant model in the mouse, combining CD28 and CD40 blockade with disruption of the OX40/OX40L pathway induced long-term skin allograft survival, which was not observed when OX40 signaling was intact. In addition, blocking OX40 costimulation is equally effective in preventing skin allograft rejection mediated by either CD4+ or CD8+ T cells. In the skin transplant model, blocking OX40 costimulation appears to be uniquely required, as blocking other alternative costimulatory pathways failed to prevent graft rejection (Demirci et al. 2004). Thus, OX40 appears to play an important but conditional role in allograft rejection. These findings also highlight the complex interactions amongst CD28, CD154, and OX40 costimulatory signals in coordinating the allograft response.

Another important aspect of OX40 costimulation is related to its role in regulating both generation and reactivation of memory T cells. Mechanistically, it remains unclear how OX40 regulates the memory response, but this may be related to the sustained expression of cell survival genes triggered by OX40 costimulation. Thus, one of the key effects of OX40 blockade in transplant models may be to inhibit transition of effector T cells into alloreactive memory T cells. OX40 blockade may also inhibit activation of preexisting memory T cells that are alloreactive in the rejection response. In presensitized recipients in which donor-specific memory T cells are present at high frequency, blocking the OX40/OX40L pathway, along with CD28 blockade, markedly prolongs heart allograft survival. Homeostatic proliferation is known to produce memory-like T cells that are resistant to conventional CD28/B7 or CD40/CD154 costimulatory blockade treatment. This is particularly relevant to transplantation where many therapeutic antibodies induce at least partial T-cell depletion. In contrast to naïve T cells, memory-like T cells resulting from homeostatic proliferation constitutively express OX40 at high levels and OX40 costimulation is critical to their activation and effector function. Indeed, blocking OX40 together with CD28 and CD154 blockade markedly extended skin allograft survival in the presence of such resistant "memory" T cells. Thus, the initial notion that activation of memory T cells is independent of costimulation is incorrect. In fact, memory T cells utilize alternative costimulatory signals, such as OX40, instead of CD28 and/or CD154. It should be noted that the effect of OX40 costimulation to Foxp3+ Treg is different from that to T effector/memory cells, and this effect will be discussed in detail bellow.

2.1.6.1 4-1BB/4-1BBL Pathway

4-1BB (CD137) is primarily expressed on activated CD8+ T cells, and to a lesser degree, on activated CD4+ T cells and activated NK cells. The ligand for 4-1BB (4-1BBL) is expressed on mature dendritic cells, activated B cells, and macrophages. Like other costimulatory molecules in the TNFR family, 4-1BB engagement by its ligand supports T-cell activation, survival, and effector differentiation. One distinct feature of the 4-1BB/4-1BBL pathway is that it preferentially stimulates CD8+ T cells. However, this effect is not absolute, and under certain conditions, 4-1BB also promotes antigen-specific CD4-cell response, and Th2 cells in particular.

Initial studies in transplant models showed that 4-1BBL-deficient mice promptly reject MHC incompatible skin allografts. Treatment of wild-type mice with a stimulating anti-4-1BB mAb accelerates rejection of both heart and skin allografts. However, in a small bowel transplant model where rejection is mediated by CD8+ T cells, stimulating 4-1BB accelerates rejection while blocking 4-1BB costimulation prolongs graft survival. In this model, extended graft survival is associated with decreased activation of allogeneic CD8+ T cells and diminished intragraft levels of chemokines (e.g., MIP-1α), confirming an important role for 4-1BB in CD8+ T-cell-mediated rejection. In a graft-versus-host model, however, 4-1BB regulates both CD4+ and CD8+ T-cell responses to the same extent, suggesting that the effect of 4-1BB can be model specific. In some settings, 4-1BB preferentially promotes CD8+ T cells in the absence of CD28 costimulation, but augments CD4-cell activation when CD28 signaling is intact. These data help explain the relative differences in costimulation of CD4+ versus CD8+ T cells by 4-1BB observed in different models. In the case of CD4+ T-cell-mediated rejection, 4-1BB plays a subservient role to CD28. However, 4-1BB blockade may be a particularly useful adjunct to CD28 blockade in preventing rejection in models where CD8+ T cells play a prominent role in rejection. In addition, some data suggest that 4-1BB may play an important role in the memory recall response. However, in contrast to OX40, 4-1BB seems to play a particularly important role in promoting CD8+ rather than CD4+ memory T cells. The implication of this finding in transplant models is unclear and certainly requires further study. Finally, Foxp3+ Treg also express 4-1BB, although little is known about the impact of 4-1BB signaling on these cells (So et al. 2008).

2.1.7 CD27/CD70 Pathway

CD27, a member of the TNFR family, is constitutively expressed on naïve T cells, B cells, and NK cells. Its ligand, CD70, is a type II transmembrane glycoprotein belonging to the TNF family. CD70 is expressed on APCs and is rapidly induced on both T and B cells upon immune activation. CD27/CD70 engagement promotes T-cell activation, and T-cell-dependent antibody production by B cells. Initial studies suggested that the CD27/CD70 pathway provides CD28-independent costimulatory signals to T cells. CD27/CD70 engagement also complements CD40 ligation in stimulating B-cell activation and antibody production (Jacquot 2000), suggesting that this pathway is important in T-B interactions. The CD27/CD70 pathway is also involved in the NK-mediated innate immunity against virus infections. Taken together, these data suggest a rather broad impact of this pathway on both innate and adaptive immunity. Interestingly, CD27 can also trigger apoptotic cell death via recruitment and activation of death pathways. In fact, transgenic expression of CD70 on B cells that constantly engages CD27 results in progressive depletion of T cells and B cells in the periphery leading to profound immunodeficiency.

The role of the CD27/CD70 pathway in alloimmunity has been extensively studied in a mouse cardiac transplant model. Blocking CD70 using anti-CD70 mAb prolonged allograft survival in wild-type recipients, and in the absence of CD28 costimulation, induced long-term graft survival, with the majority of recipients being free of chronic allograft vasculopathy. Importantly, CD70 blockade, while having little effect on CD4-mediated allograft rejection, prevented CD8-mediated rejection, inhibited proliferation and effector function of CD8 T cells, and reduced the generation of memory CD8+ T cells in vivo. Thus, the CD27/CD70 pathway appears to be critical for CD8 T-cell activation, especially in the absence CD28/B7 costimulation. Nonetheless, the role of CD27/CD70 pathway in other transplant models remains to be determined. Other areas that require further investigation include the role of CD27/CD70 in memory T-cell-mediated rejection and the role of CD27/CD70 signaling in survival, induction, and function of Foxp3+ Treg as well as the role of CD27/CD70 in the activation of innate immunity.

2.1.8 CD30/CD30L Pathway

CD30 is another member of the TNFR family and is differentially expressed by various T-cell subsets. CD30 is abundantly expressed on activated T effector/memory T cells, but not resting naïve T cells. However, Foxp3+ Treg appear to express CD30 constitutively. Its ligand, CD30L, is expressed on APCs as well as on parenchymal cells. Compared to other costimulatory pathways, the role of CD30/CD30L in regulating alloimmunity is less well studied. CD30 appears to be rapidly shed from effector/memory T cells upon activation in vivo. Thus, soluble CD30 has been extensively studied as a diagnostic marker for transplant rejection. In fact, elevated levels of soluble CD30 in the blood are closely associated with rejection of heart, lung, kidney, and liver allografts, both in animal models and in humans. In some models, CD30 may act as an important negative regulator of effector T cells, as CD30-deficient T effector cells are extremely aggressive in mediating islet destruction in an autoimmune diabetic model. In transplant settings, recent reports indicate that CD30 triggers apoptotic death of memory CD8+ T cells in immune privileged sites. Interestingly, in a cardiac transplant model, Treg can induce apoptotic death of memory CD8+ T cells via the CD30 pathway. It is likely that CD30 is critical to the establishment of the effector/memory CD4+ and CD8+ T cells by delivering costimulatory signals, these cells then shed CD30 after acquiring optimal effector functions. The therapeutic implication of this finding in the transplant setting needs to be better defined.

2.1.9 GITR/GITRL Pathway

GITR refers to *G*lucocorticoid *I*nduced *T*umor necrosis factor *R*eceptor; it is preferentially expressed on activated T effector cells, B cells, NK cells, and macrophages. Foxp3+ Treg are an exception as they constitutively express GITR on the cell surface. The ligand for GITR, GITRL, is expressed by dendritic cells and parenchymal cells including vascular endothelial cells (Watts 2005). GITR/GITRL interactions promote T-cell proliferation and cytokine production. One interesting feature of GITR is that its cytoplasmic tail can also activate signaling pathways that trigger apoptotic cell death. Thus, there is interest in manipulating the GITR/GITRL pathway for tumor immunotherapy and in autoimmune disease. The role of the GITR/GITRL pathway in transplant models has not been adequately studied. Some evidence suggests that this pathway could promote tolerance through regulation of Foxp3+ Treg. GITR signaling appears to either regulate Treg suppressor function or the sensitivity of T effector cells to Treg-mediated suppression. This is a significant issue in transplant tolerance and further study is needed to clarify the role of GITR in transplant models.

2.1.10 *HVEM/LIGHT Pathway*

Herpes virus entry mediator (HVEM) is a TNFR family member that is predominantly expressed on activated T cells. However, other cell types including B cells and NK cells are also capable of expressing HVEM. One of the ligands for HVEM is LIGHT, which is expressed by immature dendritic cells and activated T cells. HVEM/LIGHT engagement delivers a costimulatory signal that enhances T-cell activation and effector function. HVEM also binds to lymphotoxin-α (LTα). LTα is a homolog of LIGHT that is expressed by stromal cells and macrophages. Besides HVEM, LIGHT has another ligand called lymphotoxin-β receptor (LTβR). Interestingly, LIGHT binding to LTβR does not trigger costimulatory signals but instead, delivers signals that are critical for the development of secondary lymphoid organs. Furthermore, HVEM also binds to BTLA and CD160, but this engagement triggers a negative signal that inhibits T-cell activation as mentioned above.

The role of HVEM/LIGHT pathway in transplant models is less well defined. Initial studies in a bone marrow transplant model using donor cells deficient in HVEM or LIGHT showed that graft-versus-host CTL responses are completely abolished. Similarly, treatment of host mice with anti-HVEM mAb ameliorated the severity of GVHD responses and resulted in the generation of stable hematopoietic chimerism. In an islet transplant model, blocking HVEM/LIGHT interactions with LTβR-Ig together with CTLA-4Ig markedly prolonged the islet allograft survival. It seems that targeting HVEM/LIGHT pathway has the potential to prolong allograft survival.

2.2 The TIM Family Costimulatory Molecules

Molecules consisting of the *T*-cell *I*mmunoglobulin domain and *M*ucin domain are classified as TIM molecules. TIM molecules are also members in the Ig superfamily but with distinct structural features. Structurally, TIM molecules are type I transmembrane glycoproteins consisting of an extracellular domain, a transmembrane domain, and a cytoplasmic domain. Importantly, the extracellular domain of TIM molecules contains an immunoglobulin V motif and a heavily glycosylated mucin motif, a structural feature that distinguishes this family members from other costimulatory molecules. The cytoplasmic domain of TIM proteins contains a tyrosine phosphorylation motif that is involved in signal transduction upon ligand binding. The mouse TIM gene family includes eight members encoding TIM-1, TIM-2, TIM-3, TIM-4, and putative TIM-5 to TIM-8 proteins, whereas the human TIM gene family has 3 members encoding TIM-1, TIM-3, and TIM-4 proteins. The TIM molecules have attracted much attention since their discovery, in part because the TIM genes are clustered in a region of chromosome (5q32.2 in humans and 11B1.1 in mice) that is closely associated with multiple autoimmune diseases (Kuchroo et al. 2003).

TIM molecules were initially identified on differentiated CD4+ T cells and thought to be involved in regulating Th1/Th2 responses. Earlier in vitro studies suggested that TIM-1 may be critical in regulating IL-4-producing Th2 cells, whereas TIM-3 may regulate IFN-γ-producing Th1 cells. We now know that this belief is an oversimplification of the complex role of TIM molecules in vivo. It is clear that TIM molecules can regulate multiple T-cell subsets including Th17 cells and can be expressed by a variety of cell types in the innate and adaptive immune systems. For example, some TIM molecules expressed by APCs such as dendritic cells contribute to activation of APCs, which in turn promote tissue inflammation and T-cell priming. Furthermore, TIM-1 is expressed on Foxp3+ Treg and may be involved in regulating activity of Foxp3+ Treg. Finally, TIM-1 also binds to phosphatidylserine expressed on apoptotic cells, demonstrating a role for TIM-1 in clearance of apoptotic cells in the immune system. Thus, the TIM family regulates a wide array of immune-related responses. The role of individual TIM molecules in the immune system, and in transplantation in particular, is considered in details below.

TIM-1 is preferentially expressed on activated CD4+ and CD8+ T cells. The ligand for TIM-1 is TIM-4, which is constitutively expressed on APCs, but its expression is upregulated by cellular activation. A key feature of TIM-4 is that its cytoplasmic region lacks signaling motif; thus, TIM-4 itself is unable to trigger signaling events. Like other costimulatory molecules, engagement of TIM-1 by TIM-4 in the presence of TCR stimulation provides a potent costimulatory signal to T cells, enhancing proliferation, survival, and cytokine production. Furthermore, an agonist anti-TIM-1 mAb (clone 3B3) significantly enhances proliferation of T cells in response to allogeneic dendritic cells. In some culture conditions, this agonist anti-TIM-1 mAb can stimulate T-cell proliferation even in the absence of TCR stimulation. Moreover, CD4+ T cells stimulated by agonist anti-TIM-1 mAb express high levels of IL-17, suggesting a role for this molecule in the differentiation of Th17 cells. In addition, stimulation of TIM-1 on Foxp3+ Treg with agonist anti-TIM-1 abrogates their suppressor function and converts them to IL-17-producing cells, suggesting that TIM-1 may have a unique role in disarming Treg and promoting tissue inflammation. These findings suggest that TIM-1 may play a critical role in regulating the balance between T effector cells and Foxp3+ Treg, an issue that is of considerable importance in tolerance induction.

There are several reports on the role of TIM-1 in regulating transplant survival. In an islet transplant model, treatment of recipient mice with agonist anti-TIM-1 uniformly broke allograft tolerance induced by anti-CD154 treatment. Interestingly, a second mAb that binds with weaker affinity (RMT1-10) appears to interfere with TIM-1/TIM-4 interactions and treatment of recipients with this mAb combined with short course of rapamycin induced long-term allograft survival in a heart transplant model. Moreover, interference with TIM-1/TIM-4 interactions augmented Treg suppressor function without affecting Treg number. Thus, selectively targeting the TIM-1/TIM-4 pathway may be an important means of promoting transplant tolerance. However, the mechanism of targeting TIM-1/TIM-4 pathway may not be confined to T cells since some mature dendritic cells also express TIM-1. How blocking or engaging TIM-1 molecule alters dendritic cell function in vivo is not yet clear.

TIM-2 is preferentially expressed on activated CD4+ T cells, and Th2 cells in particular in mice, but TIM-2 has not been identified so far in humans. It has been shown that TIM-2 expression in mice is very dynamic and activation of CD4+ T cells under Th1 polarizing often leads to downregulation of TIM-2 expression. The putative ligand for TIM-2 is semaphorin 4A (Sema4A), a protein expressed by bone marrow-derived dendritic cells. The function of TIM-2 is poorly defined, but available data suggest that while preferentially expressed on Th2 cells, it may inhibit Th2 differentiation. For example, targeted disruption of TIM-2 exacerbates lung inflammation in a Th2-dominated model of airway inflammation. Moreover, administration of a blocking agent, TIM-2-Ig, induces Th2 responses in vivo, and in the case of EAE (a Th1-mediated autoimmune model), TIM-2-Ig delays disease progression and reduces the severity of disease. Similarly, blocking TIM-2 engagement using anti-Sema4A mAb inhibited the expression of Th1 cytokines and ameliorated the severity of EAE. These data suggest that TIM-2 may act as a negative regulator of Th2 cells. In spite of these findings, the role of TIM-2 in transplant rejection and tolerance induction awaits further investigation.

As compared to other TIM molecules, TIM-3 is primarily expressed on terminally differentiated Th1 cells, but not on resting T cells or polarized Th2 cells. The role TIM-3 in regulating T-cell responses in vivo was examined in several mouse models using an anti-TIM-3 mAb or TIM-3-Ig fusion protein. Administration of either anti-TIM-3 mAb or TIM-3-Ig consistently accelerates the development of EAE, colitis, and autoimmune diabetes. In an islet transplantation, either reagent can abrogate transplant tolerance induced by DST and anti-CD154 treatment. Furthermore, while DST plus anti-CD154 uniformly induces tolerance to islet allografts in wild-type mice, it fails to do so in TIM-3 knockout mice (Sánchez-Fueyo et al. 2003). It is concluded from these studies that TIM-3 inhibits the Th1 response and that interference with TIM-3 signaling using these reagents enhances the Th1 response and exacerbates Th1-mediated immunity, leading to more rapid disease onset and failure of allograft tolerance. This notion was further supported by the identification of galectin-9 as a TIM-3 ligand and TIM-3/galectin-9 engagement promotes apoptotic death of Th1 cells. Thus, TIM-3 appears to function as a death receptor for terminally differentiated Th1 cells.

TIM-3 may play additional roles in the regulation of both innate and adaptive immunity. One surprising finding is that TIM-3 can be expressed by CD11c+ dendritic cells and a subset of monocytes and that stimulation of TIM-3 on dendritic cells induces the production of proinflammatory cytokines such as IL-1 and TNF-α. Moreover, TIM-3 and TLR signaling synergize with one another in promoting activation of the innate immune cells (Anderson et al. 2007). Thus, TIM-3 appears to have opposing roles in the activation of innate versus adaptive immunity. It seems that, early in an immune response, TIM-3 signaling in dendritic cells may promote innate immunity and inflammation, and once a Th1 response in fully initiated, TIM-3 on differentiated Th1 cells then triggers apoptotic cell death upon binding to galectin-9, thereby terminating the Th1 response.

The finding that TIM-3 may both promote and inhibit the immune response may help explain otherwise discrepant results obtained using anti-TIM-3 and TIM-3-Ig in vivo. In the EAE model, anti-TIM-3 induced massive expansion of dendritic cells and macrophages. In this setting, anti-TIM-3 may activate dendritic cells and monocytes, which in turn promotes disease development and inhibit tolerance induction. On the other hand, TIM-3-Ig may interfere with TIM-3 signaling on activated Th1 cells, which prevents their apoptotic death, thereby interfering with tolerance induction. Thus, while anti-TIM-3 and TIM-3-Ig both block tolerance induction, they appear to selectively target TIM-3/TIM-3 L on different cell types, thus differentially affecting TIM-3 signaling (inhibition vs. augmentation). The basis for the differential TIM-3 signaling in different cell types remains to be defined. Clearly, a better understanding of this issue is therapeutically important.

2.3 Costimulatory Control of T Regulatory Cells

There are several cell types in the immune system that can act as regulatory cells, but Foxp3+ Treg are the best studied ones. Foxp3+ Treg distinguish them from others by the expression of transcription factor Foxp3 (Sakaguchi 2004). In addition to maintaining self-tolerance, Foxp3+ Treg also play an indispensable role in both the induction and maintenance of allograft tolerance (Wood and Sagaguchi 2003). Although Foxp3+ Treg represent a distinct lineage, individual Treg are not long lived, terminally differentiated, and anergic, but undergo both homeostatic and antigen-driven proliferation, they require activation for suppressive function. It has also been recognized that Foxp3 expression and regulatory function can be induced or lost depending on certain conditions. It is becoming apparent that costimulatory molecules are involved in each of these aspects of Treg function. Thus far, the costimulatory molecules expressed by Treg are shared with other (activated) CD4+ T cells. However, costimulatory signals can either promote or inhibit Treg activities, and the same molecule can have similar or diametrically opposed, actions on Treg and conventional T effector cells. In the setting of organ transplantation, an important but challenging issue is to target costimulatory pathways to inhibit T effector cells without causing inhibition of Treg in tolerance induction.

The biology of Treg and conventional T cells differs in several aspects, and a better understanding such differences will provide a window of opportunities in selectively modulating such two cell types. It has been shown that Foxp3+ Treg undergo a higher rate of homeostatic proliferation than conventional T cells. We demonstrated that as high as 50 % of Treg undergo homeostatic proliferation over a 6-day time period versus only 5 % for conventional T cells (Tang et al. 2008). Treg must undergo this high rate of proliferation to maintain steady state levels, which comprise ~10 % of the CD4+ T-cell population. This has important implications in tolerance induction by modulating costimulatory pathways. Thus, altering the rate of proliferation or survival may profoundly influence Treg numbers in the periphery. In this context, costimulation may have a dramatic effect on homeostasis of Treg than it does on conventional T effector cells. It is well documented that CD28 knockout mice have markedly reduced numbers of Foxp3+ Treg in both the thymus and the spleen, and when bred onto autoimmune-prone strains (e.g., NOD mice), these mice exhibit exacerbated and widespread autoimmunity, suggesting that CD28 costimulation is critically involved in Treg homeostasis and function. This dramatic effect of CD28 deficiency on Treg in both the thymus and the periphery was not observed with conventional T cells. Moreover, CD28 blockade with either anti-B7 mAbs or with CTLA4-Ig also inhibits Treg homeostatic proliferation and dramatically reduces Treg numbers in vivo in wild-type mice, and this occurred independently of any decrease in thymic output. While CD28 expression on Treg alone was insufficient to restore Treg numbers, IL-2 could rescue peripheral Treg homeostasis. Thus, a major effect of CD28 blockade (or CD28 knockout) on Treg homeostasis and survival is likely mediated by inhibition of IL-2 production by effector T cells. However, the possibility that CD28 may have additional roles in supporting homeostatic proliferation and TCR-mediated events required for Treg function cannot be completely excluded. In transplant patients, treatment with Belatacept (a mutated version of CTLA-4Ig) to block CD28 pathway has been shown to result in a precipitous drop in Foxp3+ Treg in blood, followed by a slow recovery between treatments (Bluestone et al. 2008). This has important implications for CTLA4-Ig-based therapy and suggests that Belatacept may be more effective as an immunosuppressive rather than tolerogenic agent. In addition, given the importance of IL-2 for Treg homeostasis, it is not surprising that calcineurin inhibitors, which shut down IL-2 production, markedly reduce Treg numbers. This effect might contribute to the antagonistic effects of calcineurin inhibitors in tolerance induction by various costimulatory blockade strategies.

As noted above, Foxp3+ Treg constitutively expresses CTLA-4, and CTLA-4 is critically involved in the suppressor function of Treg. Although the mechanisms of Treg-mediated suppression remain to be resolved, ligation of B7 molecules on APCs by CTLA-4 expressed by Treg (and activated conventional T cells) induces indoleamine 2,3-dioxygenase (IDO), which catabolizes tryptophan in the extracellular milieu to inhibit T-cell proliferation. It was suggested that CTLA-4 on Foxp3+ Treg may inhibit APCs by stripping off B7 molecules, thereby inhibiting CD28-mediated costimulation to T effector cells (Wing et al. 2008). We also demonstrated that CTLA-4 signaling also exhibits cell intrinsic inhibition of Treg activation and proliferation, parallel to its effects on conventional T cells. For example, administration of a blocking mAb against CTLA-4 dramatically augments Treg homeostatic proliferation in otherwise naïve wild-type mice, resulting in a two- to threefold increase in Foxp3+ Treg number (Tang et al. 2008). Unfortunately, the negative impact of blocking CTLA-4 on both Treg function and enhancing T effector cell activation outweighs the increase in Treg number, and therefore, anti-CTLA-4 mAb blocks transplant tolerance and promotes autoimmune disease in many models. Other approaches that might specifically enhance homeostatic or Ag-driven proliferation of Treg in vivo may be important for promoting tolerance, such approaches may be feasible once the signals governing both phenomena are better understood. This strategy would have many advantages over current attempts to expand Treg in vitro for clinical use, which has proven to be technically challenging and problematic.

Many studies demonstrate that other costimulatory molecules may also directly alter Treg function and/or differentiation. For example, we and others have shown that OX40 is expressed by both natural Treg and those induced by TGF-β. In contrast to other costimulatory molecules, OX40 has diametrically opposite effects on Treg and effector T cells. While it has a positive costimulatory function on T effector cells, OX40 is a potent negative regulator of Foxp3$^+$ Treg. Under certain conditions, stimulation of OX40 on Treg using either an agonist anti-OX40 mAb or OX40L-transgenic APCs abolished their suppressor activity both in vitro and in vivo (Vu et al. 2007). The loss of suppressor function following OX40 costimulation is not due to altered proliferation or survival of Foxp3$^+$ Treg, but appears to be associated with the suppression of Foxp3 gene expression. Moreover, OX40 costimulation of T effector cells completely inhibited the induction of new Foxp3$^+$ Treg from the activated T effector cells. Furthermore, OX40 costimulation also blocks the generation of IL-10-producing "adaptive" regulatory cells. Another key finding is that Treg homeostasis does not rely on OX40, as Treg number and their suppressor function are comparable between OX40 knockout mice and wild-type controls (Vu et al. 2007). These findings suggest that, unlike the CD28/B7 pathway, blockade of OX40 might promote tolerance by both inhibiting T effector cells and by promoting Treg activity. This makes OX40 blockade a promising target for the induction of allograft tolerance. Studies also suggest that molecules in the TIM family may also regulate Treg function. It has been reported that stimulation of Foxp3+ Treg with an agonist anti-TIM-1 mAb abrogated the suppressor function of Treg in vitro. Moreover, a significant fraction of Foxp3+ Treg became IL-17-producing cells after stimulation with an agonist anti-TIM-1 mAb. This highlights an important role of TIM-1 in the interplay between Foxp3 and Th17 cell subsets. These data indicate that TIM-1 may actually be able to regulate redifferentiation of one cell subset into the other. On the other hand, a different anti-TIM-1 mAb, RMT1-10, which binds to TIM-1 with lower affinity, appears to augment the potency of Treg. Thus, the impact of TIM-1 on Foxp3+ Treg homeostasis and function is complex, and depending how the TIM-1 is targeted on Treg, the outcome can be completely different. Finally, Foxp3+ Treg also express 4-1BB, ICOS, and GITR but their exact role in Treg function and in Treg-mediated transplant tolerance remains poorly defined.

Given the complex role of costimulatory molecules in regulating Treg and conventional T effector cells, an ideal costimulatory target will be the one that specifically augments Treg number and/or suppressor function without boosting T effector cell generation. Along the same line, it is of critical importance to target costimulatory molecules that will specifically inhibit effector T cells without inhibiting Treg homeostasis or suppressor function. Although conventional T cells can be converted to Foxp3+ Treg in vitro in the presence of TGF-β and IL-2, conversion of T effector cells in vivo, especially in an inflammatory setting, is extremely rare. It has been hoped that costimulatory blockade would promote this process, but in vivo studies have not shown clear-cut evidence that anti-CD154, CTLA4-Ig, or anti-TIM-1 enhance conversion of conventional T effector cells into Foxp3+ Treg. However, several studies show that Treg can be induced by conversion when antigen presentation occurs in settings where T-cell activation and/or APC activation is suboptimal or in the presence of anti-CD4 mAb. It should be noted that a critical barrier to the induction of Treg in vivo appears to be the activation of innate immune cells and tissue inflammation. In cases where TGF-β is present, which is required for Foxp3 induction, inflammatory cytokines, such as IL-1, IL-6, and TNF-α, are also present. Such inflammatory cytokines combined with TGF-β then divert T effector cells toward the Th17 pathway. Th17 cells are highly inflammatory; they often mediate extensive tissue damage. Thus, targeting costimulatory pathways that could control both activation of T effector cells and innate inflammatory cells might be more effective in Treg induction in vivo. This notion needs to be vigorously tested in vivo in transplant models.

Several promising approaches to enhance Treg to promote tolerance induction are now emerging. For example, OX40/OX40L costimulatory pathway is a particularly attractive target for therapeutic blockade. OX40 blockade may inhibit T effector cells and simultaneously augment Treg function. In addition, agonist anti-TIM-1 can promote "redifferentiation" of Treg toward Th17 cells, suggesting that TIM-1 blockade in inflammatory settings might augment Treg generation, especially if the antibody could act to block IL-17 production at sites of inflammation. Moreover, there may be other strategies that specifically promote Treg homeostatic or proliferation, resulting in preferential expansion of Treg over T effector cells in vivo. The fact that homeostatic proliferation of Treg is regulated differently than that of T effector cells and that some costimulatory molecules exhibit distinct effects on Treg as compared to conventional T cells suggests that research in this area may eventually lead to better therapeutic approaches. Other possibilities include development of reagents that potently inhibit T effector cells with other agents that promote Treg. The combination of these reagents may also achieve the goal of tipping the balance between T effector cells and Treg to induce transplant tolerance.

2.4 Conclusions

There is no doubt that targeting costimulatory pathways constitutes one of the most promising approaches for the induction of transplant tolerance. However, T-cell costimulation is incredibly complex and further development of therapeutic strategies to induce transplant tolerance will undoubtedly be fraught with new challenges. It is clear that costimulatory pathways regulate multiple aspects of the alloimmune response ranging from initial T-cell priming to the development of memory and regulatory T cells. It is also clear that the immune response to allografts consists of both graft-destructive and graft-protective responses, and the balance of these two components ultimately determines the outcome of an allograft (rejection versus tolerance). Thus, a clear understanding of how T-cell costimulatory molecules regulate T effector cells and Treg, and consequently the nature of the allograft response is therapeutically important in tolerance induction. As activated effector/memory T cells and Treg often express a similar set of costimulatory molecules on their cell surface, development of means that can selectively modulate them in such a way that blocking T effector/memory cells is not at the cost of Treg represents a significant challenge for the future. Furthermore, the issues as to how parenchymal expression of costimulatory ligands regulates T-cell responses to alloantigens in the transplanted grafts and whether these ligands can be therapeutically modified for the purpose of tolerance induction also need to be addressed in future experiments. Studies in these areas may eventually lead to the design of new costimulatory-based protocols that move us ever closer to the goal of transplant tolerance in the clinic.

References

Anderson AC, Anderson DE, Bregoli L, Hastings WD, Kassam N, Lei C, Chandwaskar R, Karman J, Su EW, Hirashima M, Bruce JN, Kane LP, Kuchroo VK, Hafler DA. Promotion of tissue inflammation by the immune receptor Tim-3 expressed on innate immune cells. Science. 2007;318(5853):1141–3.

Ariyan C, Salvalaggio P, Fecteau S, Deng S, Rogozinski L, Mandelbrot D, Sharpe A, Sayegh MH, Basadonna GP, Rothstein DM. Cutting Edge: Transplantation Tolerance through Enhanced CTLA-4 Expression. J Immunol. 2003;171(11):5673–7.

Bluestone JA. Is CTLA-4 a master switch for peripheral T cell tolerance? J Immunol. 1997;158: 1989–93.

Bluestone JA, Liu W, Yabu JM, Laszik ZG, Putnam A, Belingheri M, Gross DM, Townsend RM, Vincenti F. The effect of costimulatory and interleukin 2 receptor blockade on regulatory T cells in renal transplantation. Am J Transplant. 2008;8:2086–96.

Butte MJ, Keir ME, Phamduy TB, Sharpe AH, Freeman GJ. Programmed death-1 ligand 1 interacts specifically with the B7-1 costimulatory molecule to inhibit T cell responses. Immunity. 2007;27:111–22.

Cai G, Anumanthan A, Brown JA, Greenfield EA, Zhu B, Freeman GJ. CD160 inhibits activation of human CD4+ T cells through interaction with herpesvirus entry mediator. Nat Immunol. 2008;9:176–85.

Demirci G, Amanullah F, Kewalaramani R, Yagita H, Strom TB, Sayegh MH, Li XC. Critical role of OX40 in CD28 and CD154 independent rejection. J Immunol. 2004;172:1691–8.

Dong C, Juedes AE, Temann UA, Shresta S, Allison JP, Ruddle NH, Flavell RA. ICOS co-stimulatory receptor is essential for T-cell activation and function. Nature. 2001;409: 97–101.

Fecteau S, Basadonna GP, Freitas A, Ariyan C, Sayegh MH, Rothstein DM. CTLA-4 up-regulation plays a role in tolerance mediated by CD45. Nat Immunol. 2001;2:58–63.

Habicht A, Kewalaramani R, Vu MD, Demirci G, Blazar BR, Sayegh MH, Li XC. Striking dichotomy of PD-L1 and PD-L2 pathways in regulating alloreactive CD4(+) and CD8(+) T cells in vivo. Am J Transplant. 2007;7:2683–92.

Harada H, Salama AD, Sho M, Izawa A, Sandner SE, Ito T, Akiba H, Yagita H, Sharpe AH, Freeman GJ, Sayegh MH. The role of the ICOS-B7h T cell costimulatory pathway in transplantation immunity. J Clin Invest. 2003;112(2):234–43.

Jacquot S. CD27/CD70 interactions regulate T dependent B cell differentiation. Immunol Rev. 2000;21:23–30.

Khoury SJ, Sayegh MH. The roles of the new negative T cell costimulatory pathways in regulating autoimmunity. Immunity. 2004;20:529–38.

Kirk AD, Harlan DM, Armstrong NN, Davis TA, Dong Y, Gray GS, Hong X, Thomas D, Fechner JH, Knechtle SJ. CTLA-4Ig and anti-CD40L prevent renal allograft rejection in primates. Proc Natl Acad Sci U S A. 1997;94:8789–94.

Kuchroo VK, Umetsu DT, DeKruyff RH, Freeman GJ. The TIM gene family: emerging roles in immunity and disease. Nat Rev Immunol. 2003;3:454–62.

Kurtz J, Lie A, Griffith M, Eysaman S, Shaffer J, Anosova N, Turka LA, Benichou G, Sykes M. Lack of role for CsA-sensitive or Fas pathways in the tolerization of CD4 T cells via BMT and anti-CD40L. Am J Transplant. 2003;3:804–16.

Lechler R, Garden OA, Turka LA. The complementary roles of deletion and regulation in transplantation tolerance. Nat Rev Immunol. 2003;3(2):147–58.

Lechler R, Sykes M, Thomson AW, Turka LA. Organ Transplantation-how much of the promise has been realized? Nat Med. 2005;11(6):605–13.

Li XC, Strom TB, Turka LA, Wells AD. T cell death and transplantation tolerance. Immunity. 2001;14:407–16.

Markees TG, Phillips NE, Gordon EJ, Noelle RJ, Shultz LD, Mordes JP, Greiner DL, Rossini AA. Long term survival of skin allografts induced by donor splenocytes and anti-CD154 antibody in thymectomized mice requires CD4+ T cells, IFN-g, and CTLA-4. J Clin Invest. 1998;101:2446–55.

Murphy KM, Nelson CA, Sedý JR. Balancing co-stimulation and inhibition with BTLA and HVEM. Nat Rev Immunol. 2006;6:671–81.

Sánchez-Fueyo A, Tian J, Picarella D, Domenig C, Zheng XX, Sabatos CA, Manlongat N, Bender O, Kamradt T, Kuchroo VK, Gutiérrez-Ramos JC, Coyle AJ, Strom TB. Tim-3 inhibits T helper type 1-mediated auto- and alloimmune responses and promotes immunological tolerance. Nat Immunol. 2003;4:1093–101.

Sakaguchi S. Naturally arising CD4+ regulatory T cells for immunologic self-tolerance and negative control of immune responses. Annu Rev Immunol. 2004;22:531–62.

Sayegh MH, Turka LA. The role of T cell costimulatory activation pathways in transplant rejection. N Engl J Med. 1998;338:1813–21.

Sharpe AH, Freeman GJ. The B7-CD28 superfamily. Nat Rev Immunol. 2002;2:116–26.

So T, Lee SW, Croft M. Immune regulation and control of regulatory cells by OX40 and 4-1BB. Cytokine Growth Factor Rev. 2008;19:253–62.

Subudhi SK, Zhou P, Yerian LM, Chin RK, Lo JC, Anders RA, Sun Y, Chen L, Wang Y, Alegre M-L, Fu Y-X. Local expression of B7-H1 promotes organ-specific autoimmunity and transplant rejection. J Clin Invest. 2004;113(5):694–700.

Sugamura K, Ishii N, Weinberg AD. Therapeutic targeting of the effector T cell costimulatory molecule OX40. Nat Rev Immunol. 2004;4:420–31.

Suntharalingam G, Perry MR, Ward S, Brett SJ, Castello-Cortes A, Brunner MD, Panoskaltsis N. Cytokine storm in a phase 1 trial of the anti-CD28 monoclonal antibody TGN1412. N Engl J Med. 2006;355:1018–28.

Tafuri A, Shahinian A, Bladt F, Yoshinaga SK, Jordana M, Wakeham A, Boucher LM, Bouchard D, Chan VS, Duncan G, Odermatt B, Ho A, Itie A, Horan T, Whoriskey JS, Pawson T, Penninger JM, Ohashi PS, Mak TW. ICOS is essential for effective T-helper-cell responses. Nature. 2001;409:105–9.

Tang AL, Teijaro JR, Njau MN, Chandran SS, Azimzadeh A, Nadler SG, Rothstein DM, Farber DL. CTLA4 expression is an indicator and regulator of steady-state CD4+FoxP3+ T cell homeostasis. J Immunol. 2008;181(3):1806–13.

Vincenti F, Kirk AD. What's next in the pipeline. Am J Transplant. 2008;8:1972–81.

Vu MD, Xiao X, Gao W, Degauque N, Chen M, Kroemer A, Killeen N, Ishii N, Li XC. OX40 costimulation turns off Foxp3+ Tregs. Blood. 2007;110:2501–10.

Watts TH. TNF/TNFR family members in costimulation of T cell responses. Annu Rev Immunol. 2005;23:23–68.

Williams MA, Trambley J, Ha J, Adams AB, Durham MM, Rees P, Cowan SR, Pearson TC, Larsen CP. Genetic characterization of strain differences in the ability to mediate CD40/CD28-Independent rejection of skin allografts. J Immunol. 2000;165(12):6849–57.

Wing K, Onishi Y, Prieto-Martin P, Yamaguchi T, Miyara M, Fehervari Z, Nomura T, Sakaguchi S. CTLA-4 control over Foxp3+ regulatory T cell function. Science. 2008;322:271–5.

Wood KJ, Sagaguchi S. Regulatory T cells in transplantation tolerance. Nat Rev Immunol. 2003;3(3):199–210.

Xu H, Zhang X, Mannon RB, Kirk AD. Platelet-derived or soluble CD154 induces vascularized allograft rejection independent of cell-bound CD154. J Clin Invest. 2006;116:769–74.

Yang J, Popoola J, Khandwala S, Vadivel N, Vanguri V, Yuan X, Dada S, Guleria I, Tian C, Ansari MJ, Shin T, Yagita H, Azuma M, Sayegh MH, Chandraker A. Critical role of donor tissue expression of programmed death ligand-1 in regulating cardiac allograft rejection and vasculopathy. Circulation. 2008;117(5):660–9.

Yang Y, Wilson JM. CD40 ligand dependent T cell activation: requirement of B7-CD28 signaling through CD40. Science. 1996;273:1862–4.

Chapter 3
Age in Rodent Organ Transplantation

Weihua Gong

3.1 Introduction of Clinical Observation on Role of Age in Transplantation

Although age difference between donor and recipient was not associated with serum creatinine at 5 or 10 years or death-censored graft failure (Ferrari et al. 2011), different findings exhibited a significant impact of donor-recipient age mismatch on renal allograft function and worse short-term or long-term transplant outcome (Basar et al. 1999; Lim et al. 2010). A recent clinical study of 6,317 renal transplant patients manifested that engraftment of older living kidney grafts could cause an inferior allograft outcomes compared with younger living donors (Lim et al. 2010). Nevertheless, the effect of age difference between donor and recipient remains obscure and further experimental investigations were required.

3.2 Experimental Studies for Underlying Mechanism

It is common and broadly accepted that older age plays a critical role in rodent kidney transplantation, capable of impairing short-term and long-term graft function, increasing incidence of delayed graft function (DFG), altering immune responses of recipient, elevating frequency of rejection episodes, and worsening transplant outcome (Pascher et al. 2004; Reutzel-Selke et al. 2007; Melk et al. 2009). Younger donors could cause a better long-term allograft function compared with older ones (Pascher et al. 2004). In rat nonvascularized transplant model, it was found that

W. Gong, MD, PhD
Department of Surgery, The Second Affiliated Hospital of Zhejiang University School of Medicine, Hangzhou, People's Republic of China
e-mail: weihua.gong@gmail.com

W. Gong (ed.), *Rodent Transplant Medicine*,
DOI 10.1007/978-94-017-9472-5_3

better maturation and prolonged survival were achieved as younger fetal (renal, gonadal, and hepatic) tissues were transplanted into allogeneic recipients (Foglia et al. 1986). And older age of recipient irrespective of donor age is of great significance in accelerating chronic allograft deterioration, as evidenced by a significant upregulation of MHC, enhanced cellular infiltration, modification of immune responses (including expansion of total memory T-cell pool, a potentiated potency of pro-inflammatory macrophages) in a aged rat (Pascher et al. 2004). It was found that the survival rate of rat kidney allografts was determined by recipient age instead of donor age (Liu et al. 2001). Age matching of donor and recipient has been suggested for its beneficial effect on transplant outcome (Reutzel-Selke et al. 2007). And the relationships between age of donor organs and somatic cell senescence, alloimmune responses, and state of cells and tissues were required to be investigated (Reutzel-Selke et al. 2007; Melk et al. 2009).

3.3 Relationship Between Age and Alloimmune Responses

Donor and recipient age can cause age-associated alteration of immune responses (Pascher et al. 2004). In rat bilaterally nephrectomized kidney transplantation, correlation between donor age and host's immune responses were assessed. The findings demonstrated that older allograft could significantly exacerbate structural and functional deterioration, eliciting a stronger alloimmune response at early stage posttransplant (up to 1 month) including high frequency of peripheral B and T cells, enhanced T-cell alloreactivity, and an alteration of cytokine pattern. Furthermore, frequencies of intragraft B cells, dendritic cells, CD31+ cells were increased (Reutzel-Selke et al. 2007). At late stage (6 months post-transplant), difference of immune responses between young and old grafts would be diminished (Reutzel-Selke et al. 2007). Allorecognition triggering cell-mediated rejection is mediated by direct and indirect pathway of alloantigen presentation. It was found that donor-derived APCs only can reside in the rat kidney graft with weeks and then will be depleted from graft after transplant surgery. Instead, recipient's APCs will migrate into graft. Therefore, "aged" antigen-presenting cells (APCs) from donor can enhance allostimulation (the direct pathway of recognition of foreign antigens) and subsequently provoke alloimmune responses, especially during the early period after transplantation (Reutzel-Selke et al. 2007). As recipient age increased from 8, 16, to 40 weeks, creatinine clearance was reduced in parallel with more severe histologic impairment such as more remarkable cellular infiltration, vascular lesions, interstitial fibrosis, and extended glomerular sclerosis. At molecular level, intragraft expression level of transforming growth factor-beta (TGF-beta) and platelet-derived growth factor A and B chains (PDGF-AB) was highly augmented (Liu et al. 2001).

In mouse heart transplant model, donor age can impact recipient's CD4+T-cell responses including regulatory and effector T cells. In mouse cardiac transplant model, young (3 months) or old (18 months) C57BL/6 heart was engrafted into young BM12 recipient. It was found that without transplant surgery a significant increase of frequency of donor-derived leukocytes can be detected in elder nonmanipulated heart. An enhanced immune response was observed in the recipient of old graft, as evidenced by a significantly higher frequency of activated CD4+T cells and a stronger alloimmune responses (ELISPOT and mixed lymphocyte reactivity) and fewer CD4+T cells in the graft. On the immunosuppression side, frequency of regulatory T cells was systematically increased. However, these immune responses could not cause any alteration of graft morphology and graft survival. With respect to old syngeneic cardiac transplantation, donor age could not induce CD4+T-cell responses (Denecke et al. 2012).

However, different data showed that renal allograft function could not be affected by a cross-transplantation between young and senior rat kidneys, as evidenced by similar glomerulosclerosis index. However, oxidative stress differently displayed between young-senior group and senior-young group. It was found that younger donor would elicit significant lower activity of superoxide dismutase (SOD) and higher content of malondialdehyde (MDA) in the kidney grafts (Ding et al. 2013). Analysis of intragraft gene expression patterns by using microarray analysis and real-time PCR identified a difference of gene expression mainly focusing on the MAPK and insulin signaling pathways (Ding et al. 2013).

3.4 Role of Senescence of Graft

Each tissue or organ inherently per se has a limited capacity for its survival such as self-repair, self-renewal, and replication (Melk et al. 2009). As the capability for self-repair and renewal of tissue is reduced, a persistent inflammation and increased antigen recognition may occur (Reutzel-Selke et al. 2007). It was putatively hypothesized that exotic stress (such as process of transplant surgery) might weaken this finite capacity of organ's survival (Reutzel-Selke et al. 2007). It was observed that without transplantation old mouse kidney does not display morphological alteration in histology, and that senescence marker $p16^{INK4a}$ expression is apparently increased (Melk et al. 2009). A proteomic study on age-related changes in protein expression signature (PES) of kidney exhibited an increase of oxidative and proteolytic proteins and a decrease of antioxidant enzymes and glycolytic proteins in the process of renal senescence (Amelina and Cristobal 2009). Afterwards, free radical formation is promoted through the Fenton and Haber-Weiss reactions, causing oxidative damage to tissues (Kalinowski and Richardson 2005). After mouse kidney transplantation, a more rapid appearance of epithelial alteration was observed and cyclin-dependent kinase inhibitor $p16^{INK4a}$ expression was further increased in older renal allograft, however, which mildly occurred in older isograft (Melk et al. 2009). It implied that alloimmune responses might intensify age's effect in transplantation. With respect to older isograft, alloimmune responses were measured by testing cellular infiltration, cytology, gene expression of granzyme B, perforin, MHC, and IFN-gamma. No significant alteration was observed, indicating that older renal graft was not shown to be more immunogenic, but abnormal parenchymal susceptibility to stresses of transplantation and increased expression of senescence marker p6^{INK4a} (Melk et al. 2009).

Intriguingly, in a senescent mouse cardiac allograft-pinnacle transplant model, utilization of platelet-derived growth factor-AB heterodimer (PDGF-AB) + vascular endothelial growth factor (VEGF) + angiopoietin-2 (Ang-2) could rescue cardiac viability, restore senescent cardiac vascular function and integrity of the transplanted cardiac allograft. Growth factor combination of PDGF-AB and its downstream cytokines, VEGF and Ang-2 can synergistically exert an immediate antiapoptotic effect on cardioprotective effect (Xaymardan et al. 2004).

3.5 Synergistic Effect of Donor Age and Prolonged Cold/Warm Ischemia Time

It was studied that donor age could synergistically act with prolonged cold ischemia time, advancing cellular infiltration and kidney allograft deterioration of function (proteinuria and creatinine clearance) and structure in a Fischer 344 → Lewis rat model. This synergistic effect pronouncedly mounted as rat kidney donor is aged at 12 months rather than 18 months (Tullius et al. 2000). Furthermore, donor age can synergistically act with prolonged warm ischemia (from 20 to 45 min) on the host's immune responses in rat DCD (donation after cardiac death) kidney transplantation studies, in which old and young F-344 donor renal grafts were transplanted into bilaterally nephrectomized Lewis recipients (Denecke et al. 2013).

3.6 Conclusive Remarks

In experimental transplant studies, age as an independent factor should be accounted for research design and analysis. In clinic, reduced immunosuppression is used for senior-senior transplantation, but it is not the best choice. Age-adapted immunosuppressive regimens in the early stage of posttransplant surgery are suggested for personalized medicine (Pascher et al. 2004).

References

Amelina H, Cristobal S. Proteomic study on gender differences in aging kidney of mice. Proteome Sci. 2009;7:16.

Basar H, Soran A, Shapiro R, Vivas C, Scantlebury VP, Jordan ML, Gritsch HA, McCauley J, Randhawa P, Irish W, Hakala TR, Fung JJ. Renal transplantation in recipients over the age of 60: the impact of donor age. Transplantation. 1999;67(8):1191–3.

Denecke C, Ge X, Jurisch A, Kleffel S, Kim IK, Padera RF, Weiland A, Fiorina P, Pratschke J, Tullius SG. Modified CD4(+) T-cell response in recipients of old cardiac allografts. Transpl Int. 2012;25(3):328–36.

Denecke C, Yuan X, Ge X, Kim IK, Bedi D, Boenisch O, Weiland A, Jurisch A, Kotsch K, Pratschke J, Reutzel-Selke A, Tullius SG. Synergistic effects of prolonged warm ischemia and donor age on the immune response following donation after cardiac death kidney transplantation. Surgery. 2013;153(2):249–61.

Ding R, Chen X, Wu D, Wei R, Hong Q, Shi S, Yin Z, Ma L, Xie Y. Effects of aging on kidney graft function, oxidative stress and gene expression after kidney transplantation. PLoS One. 2013;8(6):e65613.

Ferrari P, Lim W, Dent H, McDonald SP. Effect of donor-recipient age difference on graft function and survival in live-donor kidney transplantation. Nephrol Dial Transplant. 2011;26(2):702–8.

Foglia RP, DiPreta J, Statter MB, Donahoe PK. Fetal allograft survival in immunocompetent recipients is age dependent and organ specific. Ann Surg. 1986;204(4):402–10.

Kalinowski DS, Richardson DR. The evolution of iron chelators for the treatment of iron overload disease and cancer. Pharmacol Rev. 2005;57(4):547–83.

Lim WH, Chang S, Chadban S, Campbell S, Dent H, Russ GR, McDonald SP. Donor-recipient age matching improves years of graft function in deceased-donor kidney transplantation. Nephrol Dial Transplant. 2010;25(9):3082–9.

Liu S, Lutz J, Antus B, Yao Y, Baik S, Illies F, Heemann U. Recipient age and weight affect chronic renal allograft rejection in rats. J Am Soc Nephrol. 2001;12(8):1742–9.

Melk A, Schmidt BM, Braun H, Vongwiwatana A, Urmson J, Zhu LF, Rayner D, Halloran PF. Effects of donor age and cell senescence on kidney allograft survival. Am J Transplant. 2009;9(1):114–23.

Pascher A, Pratschke J, Neuhaus P, Tullius SG. Modifications of immune regulations with increasing donor & recipient age. Ann Transplant. 2004;9(1):72–3.

Reutzel-Selke A, Jurisch A, Denecke C, Pascher A, Martins PN, Kessler H, Tamura A, Utku N, Pratschke J, Neuhaus P, Tullius SG. Donor age intensifies the early immune response after transplantation. Kidney Int. 2007;71(7):629–36.

Tullius SG, Reutzel-Selke A, Egermann F, Nieminen-Kelha M, Jonas S, Bechstein WO, Volk HD, Neuhaus P. Contribution of prolonged ischemia and donor age to chronic renal allograft dysfunction. J Am Soc Nephrol. 2000;11(7):1317–24.

Xaymardan M, Zheng J, Duignan I, Chin A, Holm JM, Ballard VL, Edelberg JM. Senescent impairment in synergistic cytokine pathways that provide rapid cardioprotection in the rat heart. J Exp Med. 2004;199(6):797–804.

Chapter 4
Role of Gender in Animal Transplantation

Weihua Gong

4.1 Introduction

The interesting observation of male-specific minor histocompatibility was interestingly achieved in rodent transplantation model in 1955, in which isogeneic male skin graft was rejected in the female recipient, but not for other combinations such as female-to-male, female-to-female, and male-to-male (Eichwald and Silmser 1955), implying that gender mismatch plays an important role in graft survival. The underlying mechanism was not unraveled until Dr. Scott et al. identified the mouse male-specific transplantation antigen, H-Y in 1995 (Scott et al. 1995). Various H-Y epitopes derived from intracellular proteins can cause genotypically identical graft rejection through their expressions on the cell surface with mouse H-2K^K major histocompatibility class I or II (MHC) molecules and subsequent recognitions in an MHC-restricted fashion by host H-Y-specific T lymphocytes, eliciting an anti-H-Y immunoresponse (Scott et al. 1995). Immunological destruction causes graft rejection or graft-versus-host disease (GVHD) following bone marrow transplantation. Nevertheless, the response to H-Y is in big variation among mice strains dependent largely on quantitative or qualitative differences in H-Y expression of male graft and host's responsiveness to H-Y.

Apart from the difference in specific transplantation antigen, transplant organ per se vary in size and weight, which may contribute to the influence of graft function and even graft survival. In clinic, gender mismatching of organs is a risk factor for chronic rejection of liver allograft (Gasbarrini et al. 2001), which is also likely to adversely impact cardiac transplant outcome and to increase early mortality (Ge et al. 2013). Indeed, the observation that female recipient can acutely reject minor antigen-mismatch male skin was confirmed in skin transplant model, but not heart

W. Gong, MD, PhD
Department of Surgery, The Second Affiliated Hospital of Zhejiang
University School of Medicine, Hangzhou, People's Republic of China
e-mail: weihua.gong@gmail.com

W. Gong (ed.), *Rodent Transplant Medicine*,
DOI 10.1007/978-94-017-9472-5_4

transplant model (He et al. 2004). The role of female donor has not been well defined. In rat heart transplant model, the interesting data showed that graft survival time and cellular rejection grade only could be affected by recipient's gender rather than donor's. Compared with male recipient, females had a shorter graft survival time and required higher dose of immunosuppression and rejection surveillance (Takami et al. 1995). Furthermore, it was shown that gender of recipient could impact its response to cyclosporine immunosuppression in chronic rat renal transplantation. Females have higher activity of cytochrome P450 3A4, which is responsible for cyclosporine A metabolism. This results in a difference of pharmacokinetics between females and males (Tanaka 1999). Despite routine tolerance induction for male rats, low dose of cyclosporine (1.5 mg/kg per day for 10 days) was not sufficient to prevent early graft failure in females, leading to an increased risk of acute rejection event and subsequent chronic allograft nephropathy (CAN) (Muller et al. 2008). In addition, a gender mismatch in transplantation appears between reduced mass of female nephron and increased functional demand of male host, leading to a poor graft function (Azuma et al. 1997). Similar observation was obtained from rat liver transplantation that a cumulative survival rate of female-to-male combination was significantly lower regardless of graft size and incidence of spontaneous death. The incidence of postoperative complications of female-to-male group such as biliary obstruction or confluent hepatic necrosis was relatively higher compared with male-to-male group (Gu et al. 2005). Therefore, males are frequently used for both donor and recipients (Gong et al. 2012; Xiao et al. 2012).

Indeed, female sex hormone, estradiol bears a direct antagonistic effect on the immunosuppressive activity of cyclosporine (Hirasawa and Kamada 1992). Gender-disparate donor and recipient can cause a different response of recipient to cyclosporine (CsA) immunosuppression in rat kidney transplant host, suggesting that a higher dose of CsA is required to prevent early graft failure in female recipient instead of male (Muller et al. 2008). Under the same treatment with CsA, female heart transplant mice had a significant shorter survival time compared with males (9.16 ± 0.41 days vs. 15.16 ± 1.72 days) (Zou et al. 2002). Administration of an antagonist of the estrogen receptor, Tamoxifen could significantly prolonged allograft survival in female hosts (13.16 ± 1.16 days) (Zou et al. 2002).

4.2 Underlying Mechanism of the Effect of Gender Impairing on Transplant Rejection

Graft immunological destruction involved in innate and adaptive immune cells is mainly caused by a recognition of recipient's lymphocytes of major and/or minor histocompatibility (H) antigens, which are normally expressed by transplant organ cells (Scott et al. 1997). The early report delineated the interesting phenomenon that male mouse isogeneic skin graft could be acutely rejected by genotypically identical female recipient, accounting for the male-specific transplantation antigen, H-Y. H-Y also could cause a rejection of human leukocyte antigen (HLA)-matched male graft by female host (Scott et al. 1995). Although male mice bear immunogenicity of H-Y antigen to females, other transplant models such as female-to-male, male-to-male, and female-to-female grafts could be successfully performed without rejection (Eichwald and Silmser 1955). The underlying mechanism was not unraveled until the study was undertaken by Dr. Scott et al. in 1995 (Scott et al. 1995). With respect to minor H antigens, H-Y can be recognized by T cells in an MHC-restricted fashion. Both MHC class I restricted (CTL) and class II restricted (TH) T cells are involved (Scott et al. 1997).

The H-Y epitopes originated from intracellular proteins are detected by T cells and expressed on cell surface with major histocompatibility complex (MHC) molecules. An H-Y epitope is expresses by a Y-chromosome gene, which defines a genetic basis for explaining antigenic difference between males and females (Scott et al. 1995). Nevertheless, not all mouse strains can response to H-Y epitopes, implying that quantitative or qualitative differences appear in H-Y expression in female responsiveness to H-Y or on male different tissues (Scott et al. 1997). To successfully trigger an immunoresponse of graft rejection, minor transplantation antigens are required to express at least two separate epitopes. And these separate epitopes may be encoded by separate genes (Roopenian 1992). In mice, H-Y expression is controlled by Hya gene, spermatogenesis, located on the short arm of the Y chromosome. Another important gene Smcy encodes an H-YK^k epitope in the same region, which is an octamer peptide, counterpart of human HLA B7 major histocompatibility class I molecule. It has a male-specific function (Scott et al. 1995, 1997). Y gene-encoded peptide VIKVNDTVQI, as a MHC class II-restricted H-Y epitope, is presented by HLA-DRbeta3*0301, which can be recognized by CD4+ T helper. These CD4+ T cells may subsequently facilitate dendritic cells maturation, causing a significant expansion of cytotoxic CD8+ T lymphocytes (Spierings et al. 2003). However, the status that H-Y epitope acts as a minor histocompatibility antigen remains controversial by far in the field of transplantation (Gratwohl et al. 2008).

In addition to aforementioned genetic disparity between females and males, transplant recipient's endocrine system particularly sex hormone also gets involved in the effect of gender mismatch on transplant outcome. Indeed, the difference of immune responses and renal hemodynamics exists between male and female (Muller et al. 1999). In a rat kidney transplant model, male or female F344 grafts was respectively transplanted into ovariectomized female Lewis rats, which were then treated with either testosterone, estradiol, or vehicle to clarify the beneficial effect of estradiol on the graft rejection. The findings revealed that administration of testosterone can enhance urinary protein excretion, severe vascular lesions, interstitial fibrosis, and extended glomerular sclerosis irrespective of donor gender. Conversely, administration of estradiol could improve graft function, diminish pro-inflammatory cellular infiltration, and preserve graft architecture (Muller et al. 1999). The protective effect of high estradiol level in the recipient occurred irrespective of donor gender (Muller et al. 1999). Nevertheless, in mice ischemia-reperfusion injury model, 70 % female or male liver mass was subject to 45 min of warm ischemia. As a result, a lower rate of apoptosis (weaker activity of JNK and p38alpha) and hepatocellular injury was observed in female mice. Female mice were more resistant to reduced-size liver ischemia/reperfusion in comparison to males. Exogenous administration of 17beta-estrogen significantly suppressed JNK and p38alpha activation and partially enhance liver regeneration. Administration of estrogen to male mice could improve their survival rate and prevent liver injury (Harada et al. 2001; Vilatoba et al. 2005). Furthermore, additional antiestrogen Tamoxifen treatment could exert a beneficial effect on heart allograft in mice, causing a prolongation of allograft survival in female recipients (Zou et al. 2002).

Besides, anatomic studies are supporting the effect of gender disparity on immune responses and others. First, age-matched males normally have a larger total kidney weight although no significant difference appears on basis of a correction for body surface area (Kasiske and Umen 1986). Estrogen and androgen complexes can affect many cell types' function, which possibly account for the difference of immune responses. It may contribute to poor kidney graft survival by using female donor kidneys (Csete 2008; Zukowski et al. 2011).

Indeed, experimental findings suggested that gender can also influence various metabolic processes and other responses of transplant organs. Ischemia/reperfusion is an important stage of transplantation procedure, in which female liver has a greater susceptibility to ischemia and reperfusion injury compared with male liver (Gasbarrini et al. 2001; Wittnich et al. 2004; Soric et al. 2007), as evidenced by a remarkable increase of all the parameters of oxidative stress including glutathione and malondialdehyde lactic dehydrogenase values (Gasbarrini et al. 2001) and a significant accumulation of tissue lactate and H+. Histological analysis can reveal a higher grade of intracellular acidosis and tissue injury, supporting the negative effect on transplant outcome (Wittnich et al. 2004; Soric et al. 2007).

4.3 Conclusive Remarks

Taken together, gender mismatch at least cannot make any advantages in animal transplantation. Owing to the complicated condition of female donor and recipient, selection of animals for research is cautious and taken into account.

References

Azuma H, Nadeau K, Mackenzie HS, Brenner BM, Tilney NL. Nephron mass modulates the hemodynamic, cellular, and molecular response of the rat renal allograft. Transplantation. 1997;63(4):519–28.

Csete M. Gender issues in transplantation. Anesth Analg. 2008;107(1):232–8.

Eichwald EJ, Silmser CR. Skin. Transplant Bull. 1955;2:148–9.

Gasbarrini A, Addolorato G, Di Campli C, Simoncini M, Montemagno S, Castagneto M, Padalino C, Pola P, Gasbarrini G. Gender affects reperfusion injury in rat liver. Dig Dis Sci. 2001;46(6):1305–12.

Ge F, Huang T, Yuan S, Zhou Y, Gong W. Gender issues in solid organ donation and transplantation. Ann Transplant. 2013;18:508–14.
Gong W, Thornley T, Whitcher GH, Ge F, Yuan S, Liu DJ, Balasubramanian S. Introduction of modified cervical cardiac transplant model in mice. Exp Clin Transplant. 2012;10(2):158–62.
Gratwohl A, Dohler B, Stern M, Opelz G. H-Y as a minor histocompatibility antigen in kidney transplantation: a retrospective cohort study. Lancet. 2008;372(9632):49–53.
Gu Y, Dirsch O, Dahmen U, Ji Y, He Q, Chi H, Broelsch CE. Impact of donor gender on male rat recipients of small-for-size liver grafts. Liver Transpl. 2005;11(6):669–78.
Harada H, Pavlick KP, Hines IN, Hoffman JM, Bharwani S, Gray L, Wolf RE, Grisham MB. Selected contribution: effects of gender on reduced-size liver ischemia and reperfusion injury. J Appl Physiol (1985). 2001;91(6):2816–22.
He C, Schenk S, Zhang Q, Valujskikh A, Bayer J, Fairchild RL, Heeger PS. Effects of T cell frequency and graft size on transplant outcome in mice. J Immunol. 2004;172(1):240–7.
Hirasawa K, Kamada N. Female sex hormone, estradiol, antagonizes the immunosuppressive activity of cyclosporine in rat organ transplantation. Transplant Proc. 1992;24(1):408–9.
Kasiske BL, Umen AJ. The influence of age, sex, race, and body habitus on kidney weight in humans. Arch Pathol Lab Med. 1986;110(1):55–60.
Muller V, Szabo A, Viklicky O, Gaul I, Portl S, Philipp T, Heemann UW. Sex hormones and gender-related differences: their influence on chronic renal allograft rejection. Kidney Int. 1999;55(5):2011–20.
Muller V, Szabo AJ, Erdely A, Tain YL, Baylis C. Sex differences in response to cyclosporine immunosuppression in experimental kidney transplantation. Clin Exp Pharmacol Physiol. 2008;35(5–6):574–9.
Roopenian DC. What are minor histocompatibility loci? A new look at an old question. Immunol Today. 1992;13(1):7–10.
Scott DM, Ehrmann IE, Ellis PS, Bishop CE, Agulnik AI, Simpson E, Mitchell MJ. Identification of a mouse male-specific transplantation antigen, H-Y. Nature. 1995;376(6542):695–8.
Scott DM, Ehrmann IE, Ellis PS, Chandler PR, Simpson E. Why do some females reject males? The molecular basis for male-specific graft rejection. J Mol Med (Berl). 1997;75(2):103–14.
Soric S, Belanger MP, Askin N, Wittnich C. Impact of female sex hormones on liver tissue lactic acidosis during ischemia. Transplantation. 2007;84(6):763–70.
Spierings E, Vermeulen CJ, Vogt MH, Doerner LE, Falkenburg JH, Mutis T, Goulmy E. Identification of HLA class II-restricted H-Y-specific T-helper epitope evoking CD4+ T-helper cells in H-Y-mismatched transplantation. Lancet. 2003;362(9384):610–5.
Takami H, Backer CL, Crawford SE, Zales VR, Mavroudis C. Influence of gender on allograft rejection in a rat heart transplant model. J Heart Lung Transplant. 1995;14(3):529–36.
Tanaka E. Gender-related differences in pharmacokinetics and their clinical significance. J Clin Pharm Ther. 1999;24(5):339–46.
Vilatoba M, Eckstein C, Bilbao G, Frennete L, Eckhoff DE, Contreras JL. 17beta-estradiol differentially activates mitogen-activated protein-kinases and improves survival following reperfusion injury of reduced-size liver in mice. Transplant Proc. 2005;37(1):399–403.
Wittnich C, Belanger MP, Askin N, Boscarino C, Wallen WJ. Lower liver transplant success in females: gender differences in metabolic response to global ischemia. Transplant Proc. 2004;36(5):1485–8.
Xiao X, Gong W, Demirci G, Liu W, Spoerl S, Chu X, Bishop DK, Turka LA, Li XC. New insights on OX40 in the control of T cell immunity and immune tolerance in vivo. J Immunol. 2012;188(2):892–901.
Zou Y, Steurer W, Klima G, Obrist P, Margreiter R, Brandacher G. Estradiol enhances murine cardiac allograft rejection under cyclosporin and can be antagonized by the antiestrogen tamoxifen. Transplantation. 2002;74(3):354–7. Transplantation. 74(3):289–90.
Zukowski M, Kotfis K, Biernawska J, Zegan-Baranska M, Kaczmarczyk M, Ciechanowicz A, Brykczynski M, Ostrowski M, Nikodemski T, Bohatyrewicz R. Donor-recipient gender mismatch affects early graft loss after kidney transplantation. Transplant Proc. 2011;43(8):2914–6.

Chapter 5
Role of Body Weight/Functioning Mass of Graft in Rodent Transplantation

Weihua Gong

5.1 The Difference of Growth Curves Between Rat and Mouse

Growth curve analyses of laboratory animals, such as the mouse and rat, provide a biological model for an understanding of overall growth of the body and its components. The body weight of inbred laboratory mice reaches a plateau of growth curve without obvious variance during adulthood after 9 weeks old [Growth chart of inbred mice (A) BALB/cJ (Stock number: 000651); (B) C57BL/6 J (Stock number: 000664) (data from The Jackson Laboratory Inc., USA, http://jaxmice.jax.org)]. Differently, overall body weight of rat postnatally increases in linear growth irrespective of rat strain or gender although body weight of female rat is always lower than that of male at the same age. The weight of male/female rat (Sprague-Dawley, Dark Agouti, and Lewis) is normally above 190/130 g after 9 weeks old (adulthood) according to the growth curves from Harlan Laboratories Inc., USA (http://www.harlan.com) [Growth chart of inbred rats (A) Lewis/HanHsd-Harlan Nederland; (B) Hsd: Sprague Dawley SD-Harlan Nederland (data from Harlan Laboratories Inc., USA, http://www.harlan.com)]. The linear growth type of rat will cause an apparent body weight difference and different size of transplant organs as the rats at two different stages are chosen for scientific studies. The relationship between body weight of donor or recipient and short-term, long-term isograft/allograft function has been studied by a series of experimental investigations.

W. Gong, MD, PhD
Department of Surgery, The Second Affiliated Hospital of Zhejiang University School of Medicine, Hangzhou, People's Republic of China
e-mail: weihua.gong@gmail.com

W. Gong (ed.), *Rodent Transplant Medicine*,
DOI 10.1007/978-94-017-9472-5_5

5.2 Clinical Observation of a Significance of Body Metabolic Index (BMI)

BMI = body weight (kg)/squared height (m^2). In clinic, as the living donor is much heavier than recipient and BMI mismatch appears, the risk for kidney allograft will be increased (Lin et al. 2011). Either high or low BMI is significantly associated with poor renal allograft survival. Furthermore, an elevation of BMI would increase the risk for delayed graft function (DFG), interestingly, whereas lower BMI showed a protective effect. However, BMI does not have any influence on acute rejection (Meier-Kriesche et al. 2002), and baseline BMI cannot affect clinical cardiac allograft survival (Clark et al. 2007). With respect to pediatric renal allografts, an increase of BMI was related with a shortened graft survival (Mitsnefes et al. 2002). An interesting observation showed that weight loss within first 3 months posttransplant surgery is a risk factor for a worse allograft survival (Clark et al. 2007). Mismatching kidney size and reduced renal mass are important risk factors for chronic allograft nephropathy in clinical transplantation. Conversely, a high ratio of donor kidney weight to recipient body weight in large measure can result in a better graft function and long-term transplant outcome (Kim et al. 2001).

5.3 Experimental Findings on Role of Donor Organ Weight or Body Weight

Chronic rejection, characterized by gradual loss of functioning mass of renal transplant, remains a clinical transplant hurdle to be overcome irrespective of improvement of immunosuppressive regimes. To mimic the natural progress of functioning nephron loss, extensive surgical ablation of renal mass can be used to establish animal experimental model. Inadequate numbers of functioning nephrons leads to overload of kidney graft (Mackenzie et al. 1994). Experimental findings exhibited that reduced graft mass irrespective of isogeneic or allogeneic renal transplant can accelerate the process of proteinuria formation (Heemann et al. 1994a) and has been accounted for a sensitive indicator in the progression of tubulointerstitial injury (Pascher et al. 2006). Conversely, a significant increase of nephron number of transplant recipient could slower the tempo of pathological injury of kidney (Mackenzie et al. 1994).

Since increased weight of donor kidney mainly reflects tubular rather than nephron mass, it is reasonable to observe that there was no significance of correlation between weight of donor kidney graft and renal graft injury. The increased weight of donor kidney graft can compensatively cover a reduced amount of functioning nephrons. Differently, both initial body weight of recipient and ratio of donor kidney weight/recipient's body weight were significantly associated with the development of allograft injury and chronic rejection (Liu et al. 2001). Indeed, both clinical and experimental data support that the ratio of renal graft to host's body weight or weight difference between donor and recipient is associated with incidence of delayed graft function, an increase of allograft immunogenicity, and a high risk of rejection episode (Zapletal et al. 2004; Gong et al. 2009). It was reported that body weight of recipient and ratio of weight of kidney/body at the transplant surgery day were significantly correlated with degree of sclerotic glomeruli (Liu et al. 2001). Our previous statistical study in rat exhibited that weight difference/body weight of recipient, ratio of native kidney to body weight of recipient, and ratio of graft to body weight were strongly associated with poor early graft function (EGF) ($p<0.0001$, $p<0.01$, and $p<0.05$, respectively) (Ge and Gong 2013).

5.4 Underlying Mechanisms for Effect of Body Weight/Functioning Mass of Graft

The underlying mechanisms of effect of body weight/functioning mass of graft are complicated and ill-defined, including mechanical effect (mechanical damage) and biological effect (such as macrophage infiltration and proinflammatory injury). In animal model, surgical reduction of renal mass generates a gap between supply of donor renal mass (weight of kidney graft) and functional demand of transplant host (body weight of recipient) and cause compensatory responses such as glomerular hyperfiltration, hyperfiltration-induced glomerular injury, and hypertension to meet the metabolic demands (Heemann et al. 1994a; Azuma et al. 1997; Oh et al. 2005; Kim et al. 2007). During this process, the capillary wall and renal tubule will be damaged owing to relative hypoxia and some proteins will be filtered into tubular lumen. Furthermore, proinflammatory cytokines will be produced and then exacerbate injury and cell death (Nangaku 2004; Kim et al. 2007), significantly affecting long-term transplant outcome (Heemann et al. 1994a; Azuma et al. 1997).

In the aspect of biological influence, body weight and functioning mass of transplant organs as alloantigen-independent factors were studied in various rodent experimental models. In an ablation-infarction model, reduced kidney mass decreased glomerular filtration rate and increased serum bicarbonate level in 1 week. The urinary ammonia level was significantly increased after adjusted for creatinine clearance. Histological analysis exhibited hypertrophy of principal cells and intercalated cells in the both cortical and outer medullary collecting duct (Kim et al. 2007). As the amount of retained kidney graft is less, the progress of proteinuria becomes faster in the recipient with reduced kidney mass (Heemann et al. 1994b). An insufficient nephron cannot meet the demand of recipient's metabolism such as serum creatinine, urine creatinine, and urine protein, which may lead to compensatory hyperfiltration (Oh et al. 2005). The insufficient or reduced function of graft caused an increased number of macrophages infiltrations, endothelial activation, and subsequent production of proinflammatory cytokines IL-1, IL-6, TGF-beta, TNF-alpha, platelet-derived growth factor (PDGF), adhesion molecules (ICAM-1), MHC class II, and extracellular matrix proteins (Heemann et al. 1994b, 1996;

Domanski et al. 2007). As a result, the pace of chronic rejection process was accelerated, as evidenced by formation of interstitial fibrosis, tubular atrophy, and glomerular sclerosis (Heemann et al. 1994b, 1996).

An interesting observation was achieved that recipient's body weight, rather than kidney graft weight, on the transplant day was significantly associated with allograft injury, implying a potential effect of recipient's body weight on chronic renal rejection (Liu et al. 2001). Therefore, it is great interest in studying on effect of body weight on graft function and transplant outcome. In a rat kidney transplant model, we previously observed a relationship between body weight difference of donor and recipient and early graft function. Our findings revealed an underlying molecular mechanism, demonstrating that IL-6 plays a critical role in mediating the effect of weight difference between donor and recipient on early graft function irrespective of isograft or allograft (Gong et al. 2009). High body weight difference resulted in an increased serum creatinine level, upregulated gene expressions of IL-6 and other stress-induced genes within graft, an inhibition of tubular function-associated gene expression, and tubular morphologic changes. Of note, utilization of neutralizing monoclonal anti-IL-6 antibody could rescue primary graft function and prevent renal tissue damage. The therapeutic effect of anti-IL-6 mAb was comparable to use of steroid in this model (Gong et al. 2009).

In addition, as transplant recipient becomes aged, to some extent, metabolic demand of recipient turns greater. Therefore, recipient's age as an important factor should be taken into consideration, which would be described in detail in a separate chapter (Liu et al. 2001).

References

Azuma H, Nadeau K, Mackenzie HS, Brenner BM, Tilney NL. Nephron mass modulates the hemodynamic, cellular, and molecular response of the rat renal allograft. Transplantation. 1997;63(4):519–28.

Clark AL, Knosalla C, Birks E, Loebe M, Davos CH, Tsang S, Negassa A, Yacoub M, Hetzer R, Coats AJ, Anker SD. Heart transplantation in heart failure: the prognostic importance of body mass index at time of surgery and subsequent weight changes. Eur J Heart Fail. 2007;9(8): 839–44.

Domanski L, Pawlik A, Safranow K, Jakubowska K, Dziedziejko V, Chlubek D, Rozanski J, Myslak M, Romanowski M, Sulikowski T, Sienko J, Ostrowski M, Ciechanowski K. Purine and cytokine concentrations in the renal vein of the allograft during reperfusion. Transplant Proc. 2007;39(5):1319–22.

Ge F, Gong W. Body weight difference between donor and recipient is an important affector of early graft function after renal transplantation. Transplant Proc. 2013;45(6):2171–5.

Gong W, Klopfel M, Reutzel-Selke A, Jurisch A, Vogt K, Haase S, Hoflich C, Polenz D, Gerstmayer B, Tomiuk S, Volk HD, Pascher A, Sawitzki B. High weight differences between donor and recipient affect early kidney graft function–a role for enhanced IL-6 signaling. Am J Transplant. 2009;9(8):1742–51.

Heemann UW, Azuma H, Tullius SG, Mackenzie H, Brenner BM, Tilney NL. The contribution of reduced functioning mass to chronic kidney allograft dysfunction in rats. Transplantation. 1994a;58(12):1317–22.

Heemann UW, Azuma H, Tullius SG, Schmid C, Philipp T, Tilney NL. Infections and reduced functioning kidney mass induce chronic rejection in rat kidney allografts. Clin Nephrol. 1996;46(1):34–8.

Heemann UW, Tullius SG, Azuma H, Tilney NL. The relationship between reduced functioning kidney mass and chronic rejection in rats. Transpl Int. 1994b;7 Suppl 1:S328–30.

Kim HY, Baylis C, Verlander JW, Han KH, Reungjui S, Handlogten ME, Weiner ID. Effect of reduced renal mass on renal ammonia transporter family, Rh C glycoprotein and Rh B glycoprotein, expression. Am J Physiol Renal Physiol. 2007;293(4):F1238–47.

Kim YS, Moon JI, Kim DK, Kim SI, Park K. Ratio of donor kidney weight to recipient bodyweight as an index of graft function. Lancet. 2001;357(9263):1180–1.

Lin J, McGovern ME, Brunelli SM, Gaccione P, Malek S, Tullius SG, Chandraker AK, Milford EL. Longitudinal trends and influence of BMI mismatch in living kidney donors and their recipients. Int Urol Nephrol. 2011;43(3):891–7.

Liu S, Lutz J, Antus B, Yao Y, Baik S, Illies F, Heemann U. Recipient age and weight affect chronic renal allograft rejection in rats. J Am Soc Nephrol. 2001;12(8):1742–9.

Mackenzie HS, Tullius SG, Heemann UW, Azuma H, Rennke HG, Brenner BM, Tilney NL. Nephron supply is a major determinant of long-term renal allograft outcome in rats. J Clin Invest. 1994;94(5):2148–52.

Meier-Kriesche HU, Arndorfer JA, Kaplan B. The impact of body mass index on renal transplant outcomes: a significant independent risk factor for graft failure and patient death. Transplantation. 2002;73(1):70–4.

Mitsnefes MM, Khoury P, McEnery PT. Body mass index and allograft function in pediatric renal transplantation. Pediatr Nephrol. 2002;17(7):535–9.

Nangaku M. Hypoxia and tubulointerstitial injury: a final common pathway to end-stage renal failure. Nephron Exp Nephrol. 2004;98(1):e8–12.

Oh CK, Jeon KO, Kim HJ, Kim SI, Kim YS, Pelletier SJ. Metabolic demand and renal mass supply affecting the early graft function after living donor kidney transplantation. Kidney Int. 2005;67(2):744–9.

Pascher A, Proesch S, Pratschke J, Reutzel-Selke A, Sawitzki B, Lehmann M, Tullius SG, Neuhaus P, Volk HD, Reinke P. Rat cytomegalovirus infection interferes with anti-CD4 mAb-(RIB 5/2) mediated tolerance and induces chronic allograft damage. Am J Transplant. 2006;6(9):2035–45.

Zapletal C, Lorenz MW, Woeste G, Wullstein C, Golling M, Bechstein WO. Predicting creatinine clearance by a simple formula following live-donor kidney transplantation. Transpl Int. 2004;17(9):490–4.

Chapter 6
Illustration of Laboratory Facilities and Microsurgical Instruments

Weihua Gong

6.1 Introduction

It is a certainty that a microsurgical lab will make a significant contribution to a university hospital, through which microsurgical techniques might be improved for microsurgeons and scientific research can be performed in experimental surgery. Therefore, it is of great importance in knowledge of the laboratory facilities and microsurgical instruments before surgical practice starts. To better understanding physical objects, specific illustrations are provided to make the learning clear. With respect to laboratory facilities, operation room is routinely located close to the animal care facility, which is preferentially equipped with a closet (storing materials) and a sink (washing the instruments). Establishment of such a laboratory should have the permission by the ethic commission of the hospital or university and apply for a registration in the government. Before animal training or experiments, a necessary reminder that all protocols should be proved by the institutional review committees according to the animal protection rules should be taken (Martins and Montero 2006).

In this chapter, we describe the necessities of animal training course and provide specific illustrations on laboratory equipments and various microsurgical instruments for better understanding. For instance, surgical board, a bench microscope, needle holder, and other instruments will be illustrated.

W. Gong, MD, PhD
Department of Surgery, The Second Affiliated Hospital of Zhejiang University School of Medicine, Hangzhou, People's Republic of China
e-mail: weihua.gong@gmail.com

W. Gong (ed.), *Rodent Transplant Medicine*,
DOI 10.1007/978-94-017-9472-5_6

6.2 Animal Training Program

To reduce anxiety and enhance confidence, animal training program is necessary to be provided irrespective of senior surgeons or residents. In terms of establishment of rodent transplant models, they will have the same starting point for deftness of microsurgical techniques (Pichierri et al. 2009). Ethical issues, drugs (such as anesthetics and analgesia), instrumentation, and postoperative healthcare of the rodents will be introduced in the training program.

6.3 Certificate of Animal Training Program

After a series of courses on "Laboratory Animal Science and Methods of Animal Experimentation" are successfully completed, Society of Laboratory Animal Science will confer a certification, which is validate throughout the Europe (Fig. 6.1). The contents of courses are in accordance with the guidelines of the Federation of European Laboratory Animal Science Association (FELASA).

6.4 Anesthetics

Anesthesia is different from analgesia (i.e., Novaminsulfon). General anesthetics for rodents include two types of administration, injection and inhalation (Ge and Gong 2011). Intramuscular [IM] or intraperitoneal [IP] injection with anesthetics (phenobarbital, zolpidem tartrate, and xylazine chlorohydrate) are normally given for small surgeries. Combined use of ketamine (60 mg/kg) (Butler Animal Health Supply, Dublin, OH, USA) and xylazine (10 mg/kg) (IP) (Akorn Inc., Decatur, IL, USA) is commonly used to anesthetize murines, for instance, mouse heart or kidney transplantation (Ge and Gong 2011) (Fig. 6.2). Inhalation of anesthetics (i.e. Isoflurane, Fig. 6.3) is preferable in establishing rodent transplant models although disadvantages exist including a requirement of expensive equipment and difficulty in movement of surgical board during operation. The gas vaporizer for Isoflurane renders anesthesia and surgeries to be controllable. Once supply of anesthetics inhalation is stopped, rodents will recover from surgeries soon. The fact that inhalation of anesthetics can offer partial protective effect on grafts must be taken into consideration.

6.5 Microsurgical Instruments

Microsurgical instruments for rodent transplant surgeries are normally fine and delicate. With respect to mouse transplantation models, higher demanding and meticulous manipulations are required since the size of mouse tissues and organs is smaller. The quality of instruments is also higher. Ideally, every microsurgeon should have

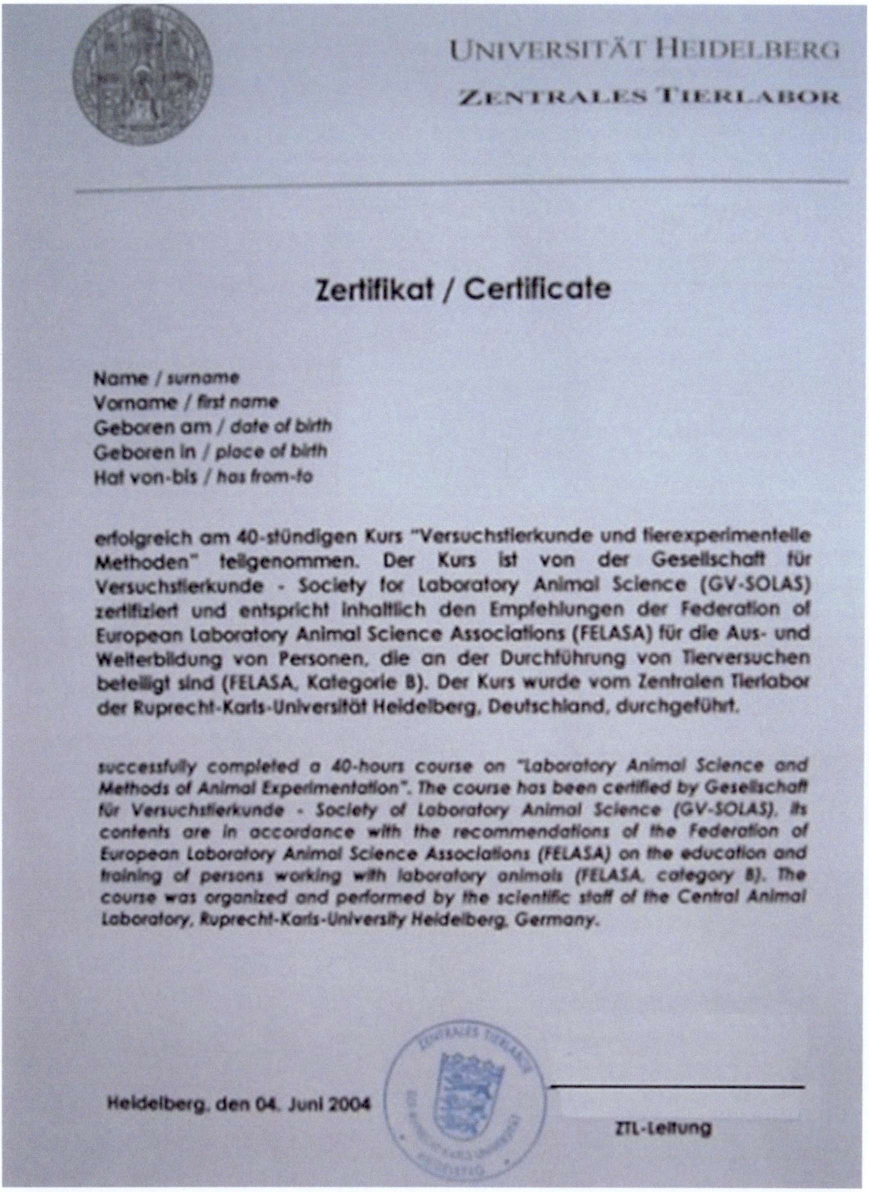

UNIVERSITÄT HEIDELBERG

ZENTRALES TIERLABOR

Zertifikat / Certificate

Name / *surname*
Vorname / *first name*
Geboren am / *date of birth*
Geboren in / *place of birth*
Hat von-bis / *has from-to*

erfolgreich am 40-stündigen Kurs "Versuchstierkunde und tierexperimentelle Methoden" teilgenommen. Der Kurs ist von der Gesellschaft für Versuchstierkunde - Society for Laboratory Animal Science (GV-SOLAS) zertifiziert und entspricht inhaltlich den Empfehlungen der Federation of European Laboratory Animal Science Associations (FELASA) für die Aus- und Weiterbildung von Personen, die an der Durchführung von Tierversuchen beteiligt sind (FELASA, Kategorie B). Der Kurs wurde vom Zentralen Tierlabor der Ruprecht-Karls-Universität Heidelberg, Deutschland, durchgeführt.

successfully completed a 40-hours course on "Laboratory Animal Science and Methods of Animal Experimentation". The course has been certified by Gesellschaft für Versuchstierkunde - Society of Laboratory Animal Science (GV-SOLAS). Its contents are in accordance with the recommendations of the Federation of European Laboratory Animal Science Associations (FELASA) on the education and training of persons working with laboratory animals (FELASA, category B). The course was organized and performed by the scientific staff of the Central Animal Laboratory, Ruprecht-Karls-University Heidelberg, Germany.

Heidelberg, den 04. Juni 2004

ZTL-Leitung

Fig. 6.1 A certificate issued by the Central Animal Laboratory, Ruprecht-Karls-University Heidelberg, Germany

his own instruments, which will keep his familiarity with them. Microsurgical skills will be more easily improved under the same settings. The basic microsurgical instruments include retractors, fine forceps, a microneedle holder, dissection micro-scissors, vascular microclamps (microserrefines and bulldogs), clamp applicator, catheters, and plastic stents (Fig. 6.4).

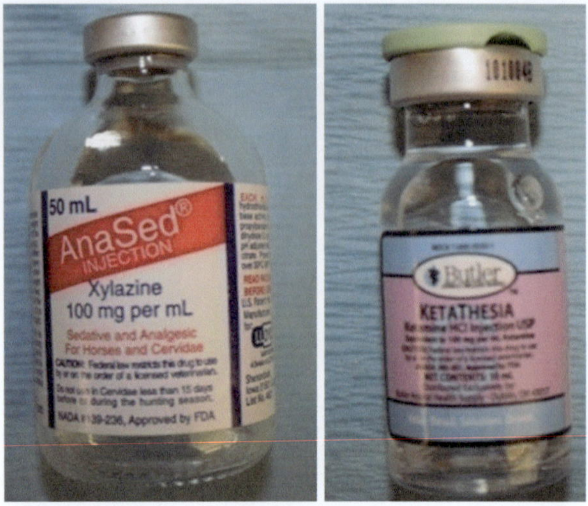

Fig. 6.2 General anesthetics (xylazine chlorohydrate and ketamine) by injection for rodent transplantation

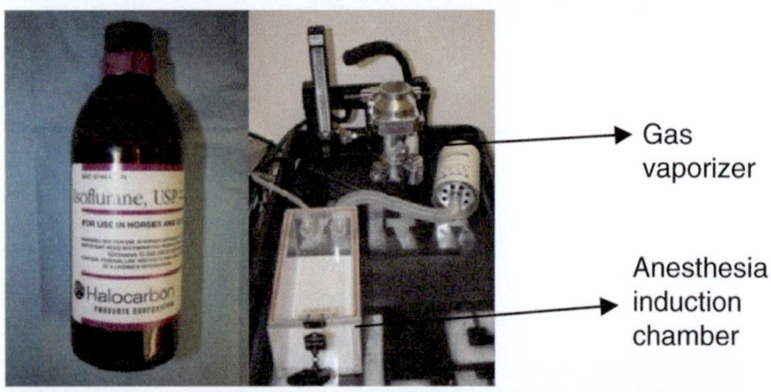

Fig. 6.3 General anesthetics by inhalation (with Isoflurane) for rodent transplantation

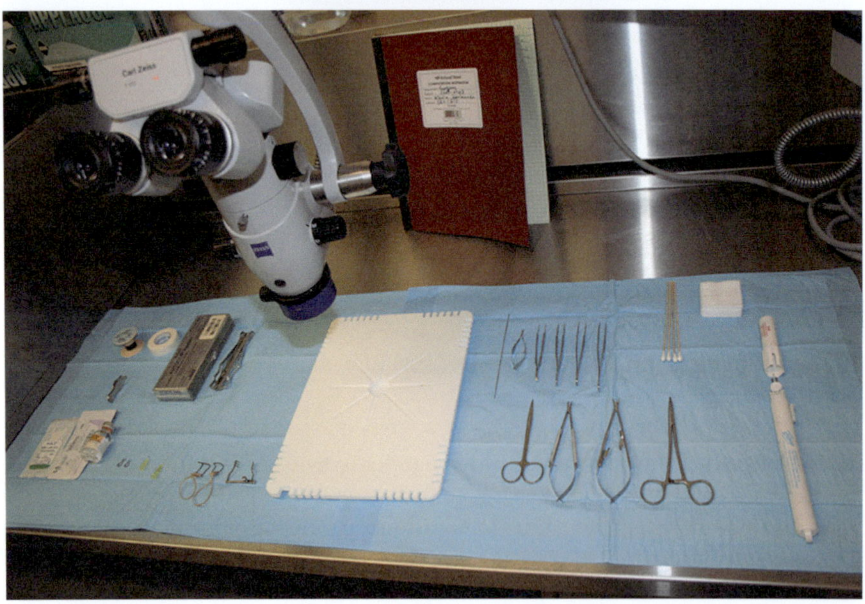

Fig. 6.4 Operation table, operating microscope, and microsurgical instruments

6.5.1 *Operating Microscope*

Operating microscope is preferentially located in the animal care facility, which should be fixed to avoid any vibrations. Figure 6.4 shows a desktop operating microscope (Carl Zeiss Operating Microscope, f170, Opmi pico, Germany) with adjustable focus and magnifying lens (to ×40).

6.5.2 *Bipolar Coagulator*

With respect to rat transplant models, bipolar coagulator (Fig. 6.5) is needed for hemostasis although ligatures may be performed as a replacement. Use of the bipolar coagulator can save operation time. However, it also causes a couple of side effects, such as occurrence of vascular spasm, stickiness of coagulator to surrounding target tissues when removed.

6.5.3 *Low-Temperature Cautery*

A low-temperature cautery (World Precision Instruments) may be utilized to coagulate small vessels in murine transplant models. The head of cautery is replaceable (Fig. 6.6). For instance, a few lumbar branches can be cauterized prior to venotomy and aortotomy in abdominal mouse heart transplantation. The end of periureteral nourishing vessels may also be cauterized to prevent hematuria in mouse kidney transplantation. Our experimental studies have proved that cauterization is feasible and safe to use (Ge and Gong 2011).

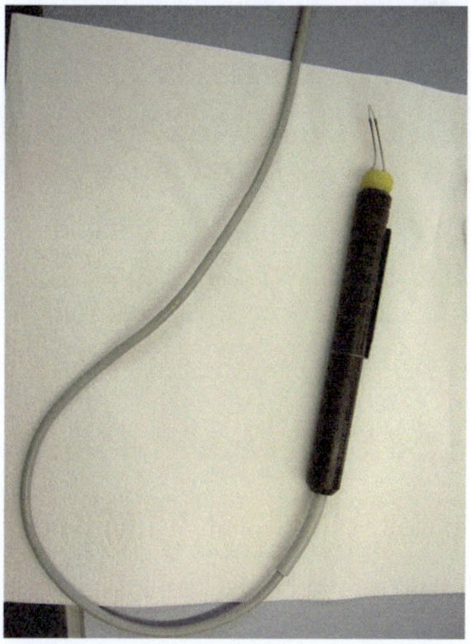

Fig. 6.5 Bipolar coagulator for stopping bleeding

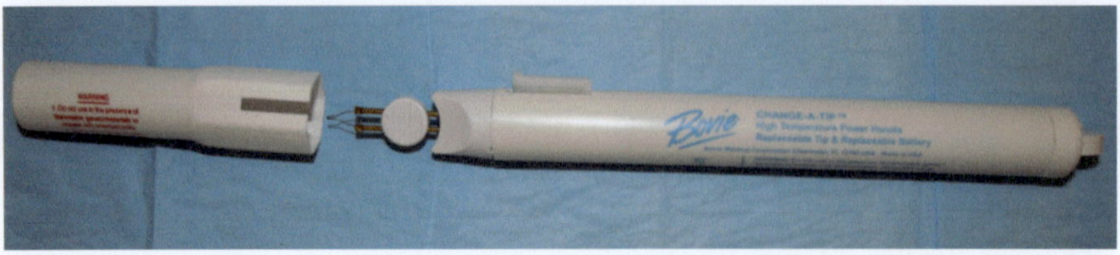

Fig. 6.6 Cautery for coagulating small vessels

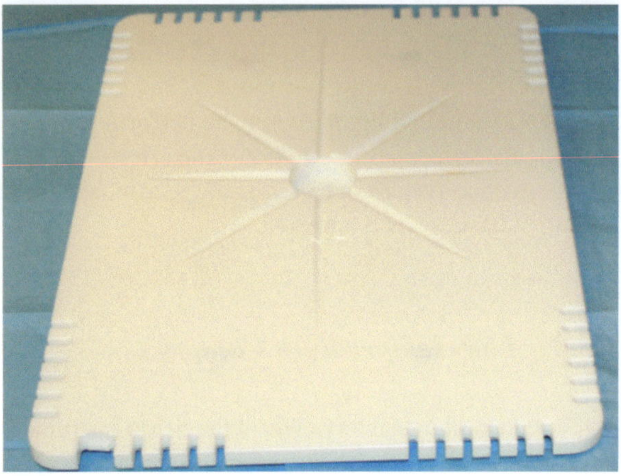

Fig. 6.7 Rodent operating board used for positioning experimental rodent

6.5.4 Rodent Operating Board

Rodent operating board (Fig. 6.7) chosen for small animal surgical work should be a one-piece construction steam sterilizable, stain resistant, easy portable, and extremely durable. It can provide a stable platform for all kinds of rodent surgeries.

6.5.5 Dual Clasping Stainless Steel Sterilization Container

The steel sterilization container (Fig. 6.8) is used for long-lasting protection, storage, and sterilization of fine surgical instruments. It should be suitable for ethylene oxide or autoclave sterilization.

6.5.6 Angiocatheter and Bent 30-Gauge Needle

The angiocatheter (Fig. 6.9) may be used to rinse the inside of vessels and clean intraluminal blood clots during venotomy and arteriotomy. Its soft head may prevent endothelial cells from physical injuries. In addition, it can be utilized to perfuse

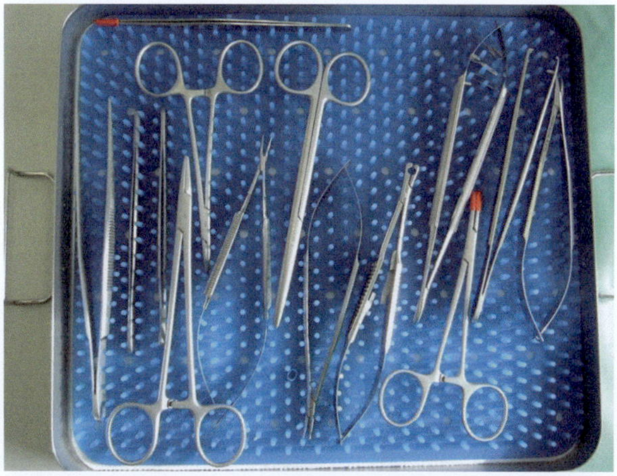

Fig. 6.8 Dual clasping stainless steel sterilization container with microsurgical instruments

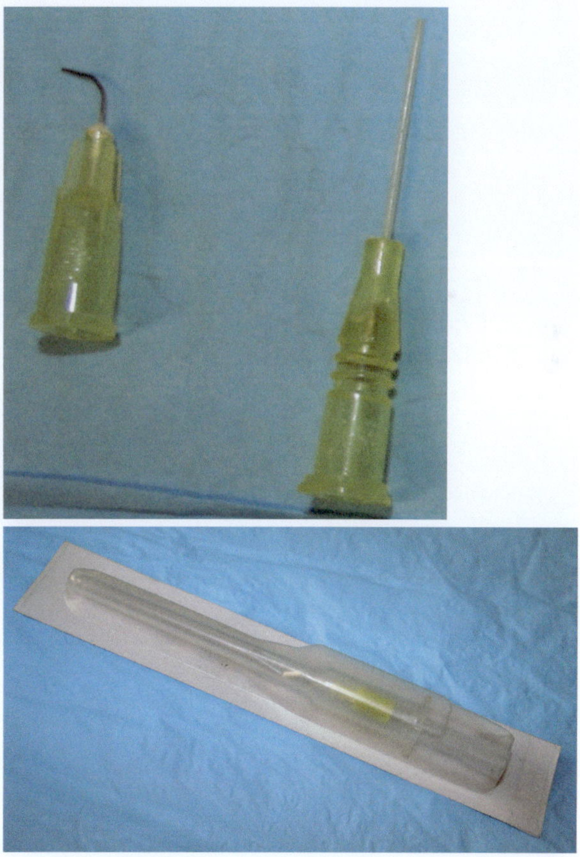

Fig. 6.9 Bent 30-gauge needle (*left*) and angiocatheter with/without package

mouse transplant grafts such as heart or kidney. Before use, the head of bent 30-gauge needle is required to be cut and polished into a sleek one. Thereafter, the bent gauge needle can be utilized to flush donor organ's vessels particularly murine donor veins. The flushed open lumen may benefit microsurgeon from identifying both ends of vessels and facilitating vascular anastomosis. Furthermore, it prevents kinking of vessels.

6.5.7 Retractors

Self-made paper clip retractors (middle and left) or other retractors manufactured in-house (right) (Fig. 6.10) can be used to expose the surgical field. However, it should be autoclaved or disinfected by alcohol before use.

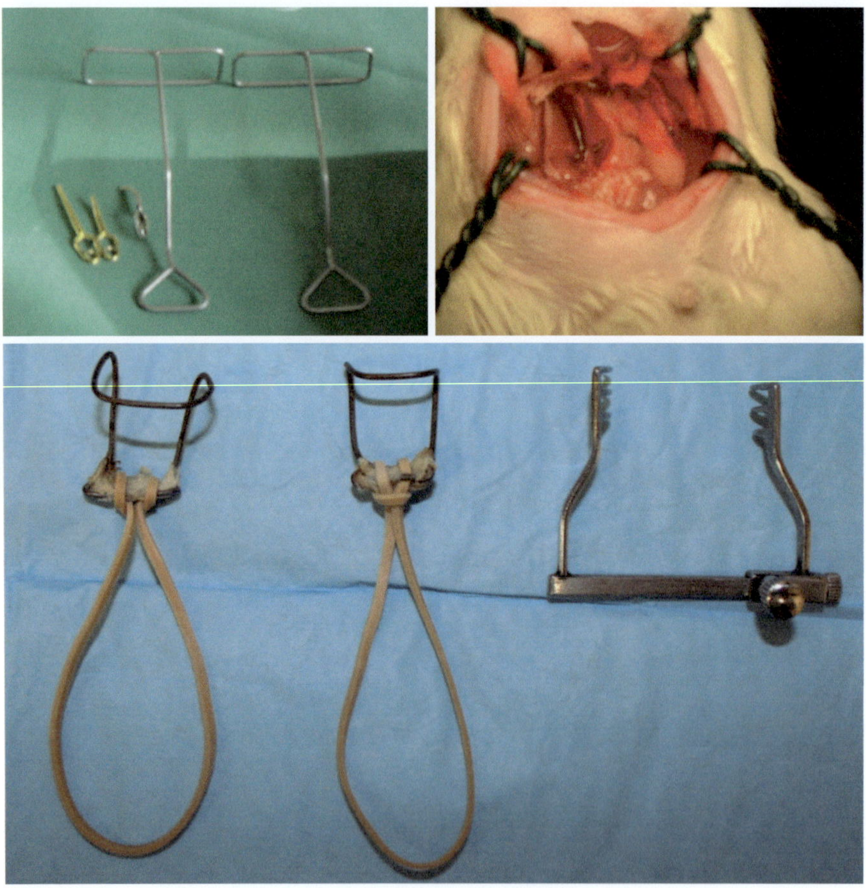

Fig. 6.10 Self-made retractors, commercial retractor, and their practical use (*middle*)

6.5.8 Microvascular Clamps

As shown in Fig. 6.11, microvascular clamps are normally applied to shortly block blood flow for subsequent vascular reconstruction although silk slip-ties may replace them. Compared with slip-tie, microvascular clamps are advantageous in exposing anastomosing opening, however, which occupying partial surgical space.

6.5.9 Curved Needle Holder

Curved needle holder without lock (Fig. 6.12) is used most commonly in rodent transplant surgeries. The curved tip can permit easier knot making and improved manipulation of the needle without bending compared with a straight one.

6.5.10 Microclip Vessel Applier with Lock

The microclip vessel applier with lock (Fig. 6.12) can be used to easily position surgical clips on the vessels.

6.5.11 Sterile Swab (Cotton Fioc) and Gauze

Sterile swab (Fig. 6.13) is useful in rodent transplantation, which can be utilized to move visceral organs for surgical area exposure. Besides, it may dissect connective tissues. As vascular anastomosis is made and microvascular clamps are released, suitable press with sterile swab may be given at anastomotic site to avoid bleeding.

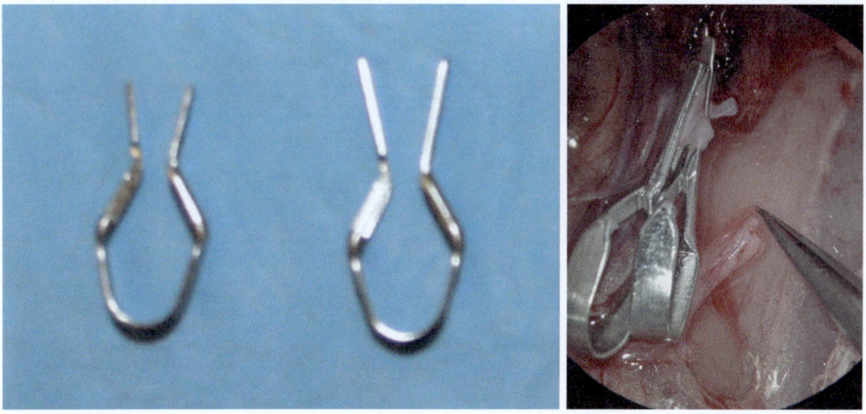

Fig. 6.11 Microvascular clamps and their practical use (*right*)

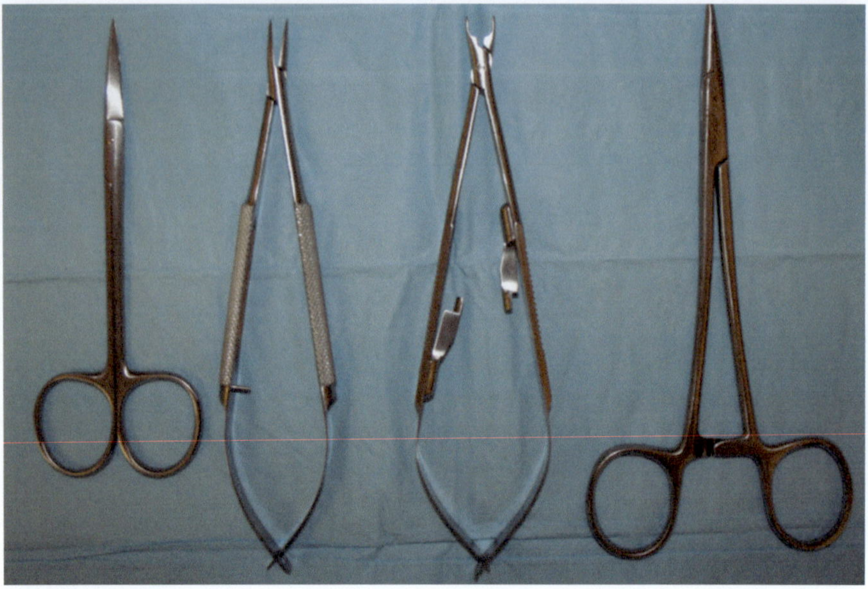

Fig. 6.12 Straight scissor, curved needle holder without lock, microclip applicator with lock, and straight needle holder (from *left to right*)

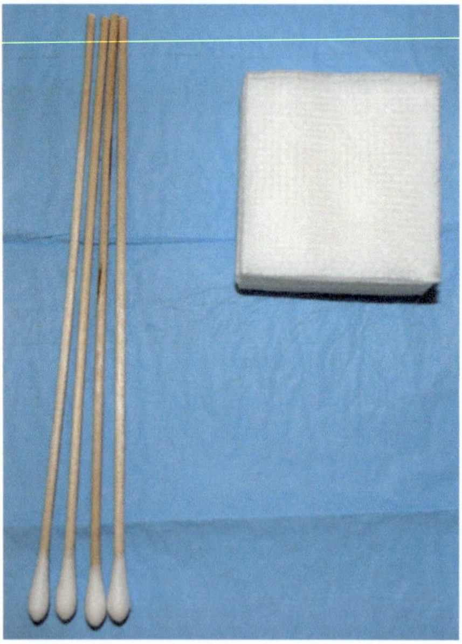

Fig. 6.13 Sterile swab and gauze used for various rodent transplant surgeries

6.5.12 Microsurgical Scissor and Microdissecting Forceps Fine Point

Microsurgical instruments (Fig. 6.14) including straight supergrip forceps and microdissecting scissors are essential for murine transplantations. These instruments are relatively delicate and care should be taken to protect them for their storage and transport. A separate tray or dual clasping stainless steel sterilization container may be employed to store. Meticulous microsurgical manipulations are required to use them during operation. After operation, an ultrasonic washing system can be utilized to efficiently improve the cleaning of microsurgical instruments and prolong their working life.

6.5.13 Needle and Sutures

10-0 suture is normally adopted to reconstruct mouse vessels and 5-0 thread for abdominal closure (Fig. 6.15). Indeed, the sutures used for microsurgery are usually monofilamentar and nonabsorbable. Cylindrical (atraumatic) needles are recommended for suturing vessels, lymphatics, and nerves, whereas sharp needle is preferred for skin stitch. Different size of vessels requires various sutures for anastomosis. Mouse vascular diameter (around 0.4 mm) is 20-fold smaller than that of rat (Ge and Gong 2011). 8-0, 9-0, or 10-0 suture is used for various rat transplantations, whereas 10-0 (needle of 100 μm and thread of 22 μm) or 11-0 (needle of

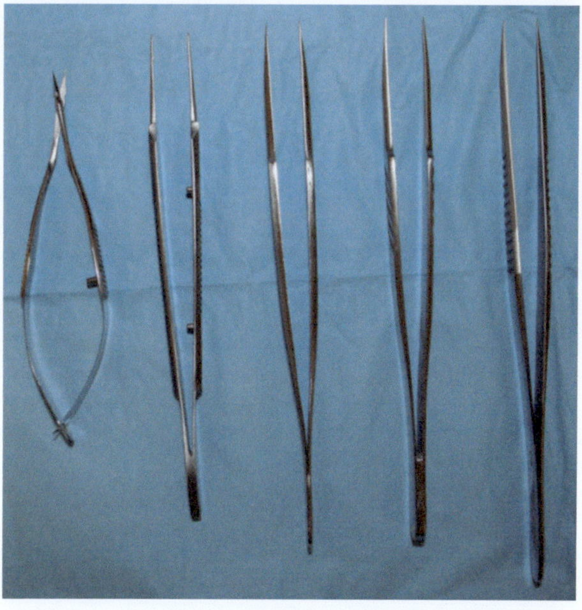

Fig. 6.14 Straight microsurgical scissor (*left*) and various microdissecting forceps fine point

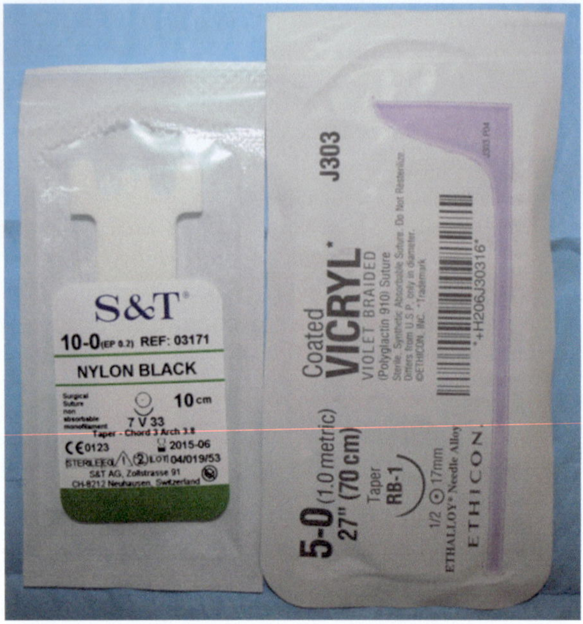

Fig. 6.15 10-0 nonabsorbable suture (*left*) and 5-0 absorbable suture (*right*)

75 μm and thread of 18 mm) thread is recommended for mouse. Utilization of suitable threads including quality and diameter of threads in a large measure plays a critical role in the quality and speed of vascular anastomosis.

6.5.14 6-0 Silk

Threads include 3-0 monofilament silk for muscle and skin. 5-0 and 6.0 monofilament threads can be used to ligate vessels (Fig. 6.16). Occasionally, 8-0 or 10-0 thread is used for suture on vessels, for instance, ligation of lumbar vein in mouse abdominal heart transplantation.

6.5.15 Surgifoam

Indeed, there are a couple of hemostatic methods, for instance, electrical (bipolar and monopolar), mechanical (pressure with swab), and chemical hemostatic agents (Gelfoam, microfibrillar collagen, or Surgifoam). Sometimes, chemicals are preferable to use by placing it around the anastomotic sites in transplant surgery before microvascular clamps or ties are released (Ge and Gong 2011). Compared to cautery, the Surgifoam (Fig. 6.17) is characterized by its obvious advantages including

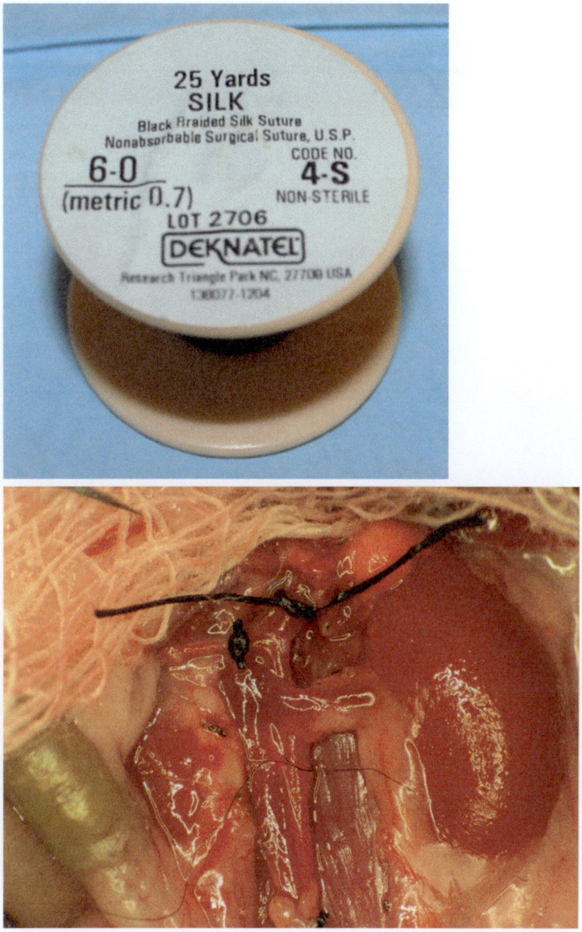

Fig. 6.16 6-0 monofilament silk and its practical use (*right*)

no blockade of vascular lumen and no injuries to adjacent structures during the course of bleeding control (Sabel and Stummer 2004).

6.5.16 Autoclip Wound Clip Applier (9 mm)

Wound closure as an effective and fast alternative solution may be accomplished by an Autoclip Wound Closing System (Fig. 6.18) apart from suturing, enabling researchers to streamline surgeries. In particular, this Autoclip Wound Closing System is indispensable for mouse thymectomy. Indeed, it is like a staple gun for use on rodent surgeries. The 9-mm autoclips are optimum for use in rats, whereas the 7-mm autoclips are suitable for mice or young rats. All components can be autoclaved before use.

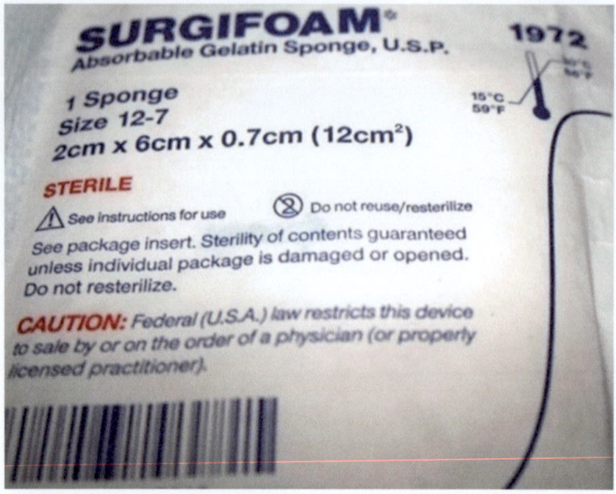

Fig. 6.17 Surgifoam as chemical hemostatic agent for hemostasis

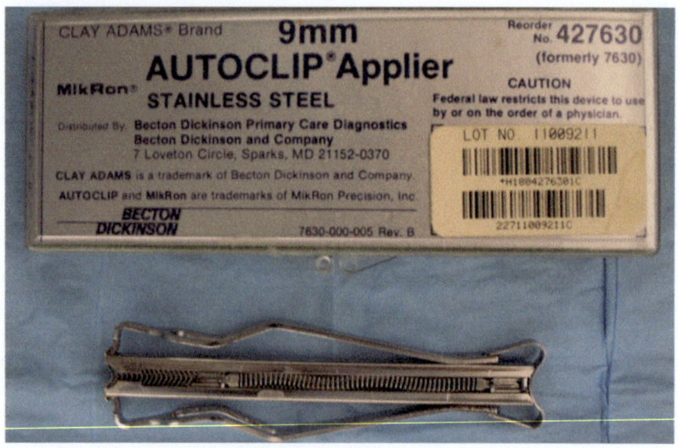

Fig. 6.18 Autoclip wound clip and its applier

6.5.17 Circulating Water Heating Pad and Pump

In the system shown in Fig. 6.19, warm water is circulated throughout the pad by an electrical pump. During or after surgical procedures, rodent or rodent's cage may be laid on heating pad to prevent body heat loss or keep rodent warm by circulating warm water through it.

6.5.18 Metabolic Cage

Metabolic cages (Fig. 6.20) are designed for separate collection of rodents' feces and urine in order to analyze administered compounds and/or their metabolites. In this system, contamination of water and food may be avoided and actual water intake of rodents can be calibrated as watering and feeding compartments are cleverly placed outside of the cage. Metabolic cages also can be used to determine drug excretion route.

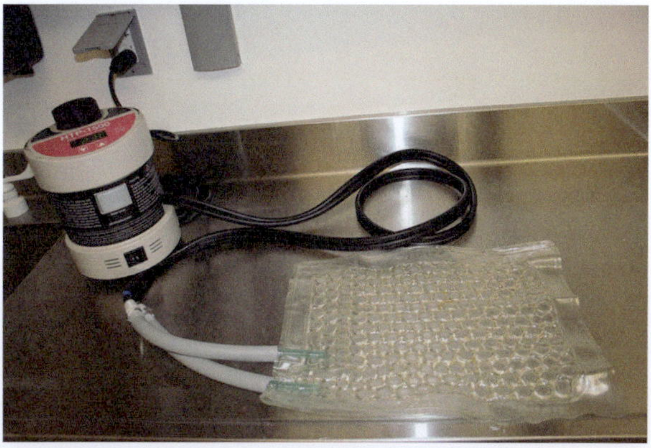

Fig. 6.19 Circulating water heating and its pump used for keeping rodents warm

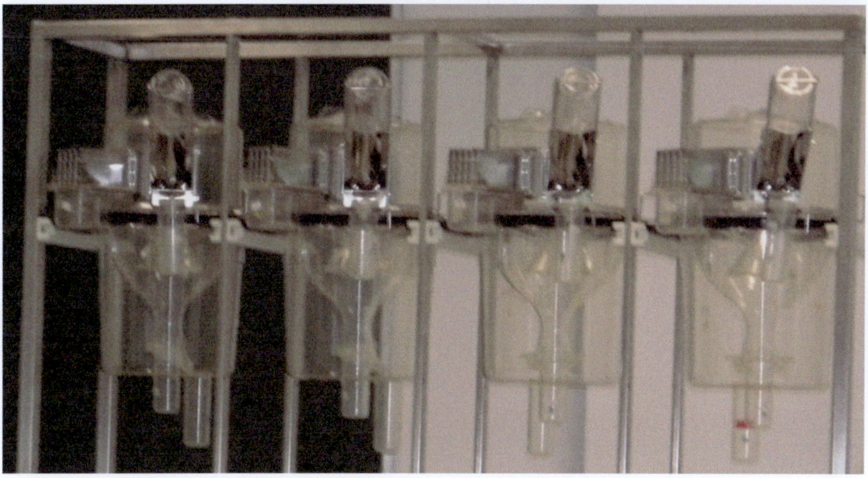

Fig. 6.20 Metabolic cage used for collecting rodents' feces and urine

Other accessories also include drapes, tray, gloves, gauze, rinse, disinfectant, glass bead sterilizer/autoclave, and circulating heating water pad/heating lamp.

References

Ge F, Gong W. Strategies for successfully establishing a kidney transplant in a mouse model. Exp Clin Transplant. 2011;9(5):287–94.

Martins PN, Montero EF. Organization of a microsurgery laboratory. Acta Cir Bras. 2006;21(3):187–9.

Pichierri A, Frati A, Santoro A, Lenzi J, Delfini R, Pannarale L, Gaudio E, D'Andrea G, Cantore GP. How to set up a microsurgical laboratory on small animal models: organization, techniques, and impact on residency training. Neurosurg Rev. 2009;32(1):101–10; discussion 110.

Sabel M, Stummer W. The use of local agents: surgicel and surgifoam. Eur Spine J. 2004;13 Suppl 1:S97–101.

Chapter 7
Organ Preservation

Weihua Gong and Dahai Liu

Organ preservation plays an important role in the field of transplantation as it can minimize the incidence of primary delayed graft function, graft nonfunction, and organ rejection episode. Currently, the amount of patients on waiting list has greatly outnumbered available donors. Full utilization of potential donor organs particularly marginal ones has become paramount such as older donors and donors after cardiac death (DCD) or brain death (DBD). And optimization of organ preservation is one of efficient strategies to alleviate this donor crisis and subsequently improve graft function (Guibert et al. 2011).

The goal of organ preservation is to improve graft outcome through minimizing the injury of organ and tissue and protecting graft function during storage. The preservation solution may prevent cell deterioration in the graft and the degenerative effect of hypothermia including cell swelling and enzyme leakage (Guibert et al. 2011) and delay hypoxia-induced ATP hydrolysis. Organ preservation strategies can at least decelerate injurious processes including hypothermia-induced damage and to some extent expand donor pool. Simple cold storage (CS) is frequently used as static preservation method in daily clinical practice, by which donor organ is perfused and stored at hypothermic temperature. CS can rapidly wash out blood components through vessels and render donor organ cool. Under this condition, oxygen is not provided and hypothermia is applied to minimize any metabolic and catabolic activities in the graft by hypothermia (0–4 °C). It was estimated that at 4 °C 10–12 % metabolism remained (Southard and Belzer 1995).

W. Gong, MD, PhD (✉)
Department of Surgery, The Second Affiliated Hospital of Zhejiang University School of Medicine, Hangzhou, People's Republic of China
e-mail: weihua.gong@gmail.com

D. Liu, PhD
Department of Biotechnology, Life Science College, Anhui University, Hefei, People's Republic of China
e-mail: dliu2009@gmail.com

W. Gong (ed.), *Rodent Transplant Medicine*,
DOI 10.1007/978-94-017-9472-5_7

The technology for preserving donor organs has considerably improved by far. The overall methods of preservation include simple cold storage (static preservation), normothermic machine perfusion, hypothermic machine perfusion (mouse model (Blanchard and Pollak 1985)), subzero nonfreezing preservation (Berendsen et al. 2014), and oxygen persufflation (Lee and Mangino 2009). Indeed, the graft outcome postvascularization determines whether the organ preservation solution is optimal.

This chapter briefly overviews research progresses on preservation of rodent organs. In rodent transplant models, normal saline, lactate Ringer's solution, HTK (Histidine–Tryptophan–Ketoglutarate), and UW (University of Wisconsin) solutions are commonly used for scientific studies.

UW solution was formulated by Belzer and Southard, which originally designed for pancreatic preservation. It characterized by a low osmolarity contains colloid and free radical scavengers (adenosine, glutathione, and allopurinol) (Avolio et al. 2006). Currently, it enables safe cold perfusion and preservation for multiple organs such as liver, kidney, heart, and pancreas (Churchill and Kneteman 1998).

Compared with UW solution, HTK derived from a nonhigh potassium cardioplegic solution is superior in the following aspects. Histidine of HTK substitutes phosphate in the UW solution. Tryptophan and ketoglutarate as substrates for intermediary metabolism can protect cellular membrane. HTK was originally developed for cardioplegia by Bretschneider HJ in 1960s in Germany (Bretschneider 1980). And it is cheaper compared with UW solution and used for preserving liver, kidney, pancreas, and heart in clinical practice (Barlow et al. 2013).

In 1969, G.M. Collins' solution was developed to prolong preservation time till 24 h, in which high concentration of glucose and intracellular ionic composition was given to reduce cellular swelling (Collins et al. 1969). In 1986, scientists (Folkert Belzer M.D., and James Southard, Ph.D.) from University of Wisconsin invented the gold standard for organ preservation techniques, an organ-preserving fluid, leading to a great improvement of cardiac, liver, lung, and other organ transplantations (Belzer and Southard 1988). The formulation of UW solution contains cell impermeant (raffinose, lactobionic acid, and anions like phosphate and sulfate), which prevents organ from the hypothermia-induced cellular swelling. Hydroxyethyl starch as a colloid replacement for albumin is utilized to reduce edema formation. By using ribose, adenine, and phosphate, adenine nucleotides are synthesized in the process of perfusion preservation. Furthermore, application of free radical scavengers (allopurinol and reduced glutathione) facilitates to remove oxidative stress (Belzer and Southard 1988).

In rodent transplantation, scientists' efforts mainly focus on static preservation. A plethora of ingredients have been supplemented to UW solution to further improve preservation conditions. For instance, trophic factor supplemented UW solution can significantly inhibit intimal inflammation and reduce intimal hyperplasia in a rat orthotopic aortic transplant model (Schmiedt et al. 2007). A caspase inhibitor, IDN-6556, added to UW cold storage solution, was showed to effectively prevent cold-ischemia–warm-reperfusion-induced sinusoidal endothelial cell injury in rat liver transplantation (Natori et al. 2003). In addition, mixture therapy with

S-adenosylmethionine (SAM) and UW solution is capable of improving hepatic blood flow during the first 3 h perfusion and reducing cold or warm ischemia/reperfusion injuries in the rat liver transplantation (Dunne et al. 1997). Exogenous fructose-1,6-biphosphate (F16BP) added to UW solution displays a protective effect during experimental intestinal transplantation in rats, as evidenced by a significant decrease of intestinal damage and an incidence of bacterial translocation. The underlying mechanisms are involved in a reduction of availability of xanthine oxidase substrates, which is used for free radical generation during reperfusion. Furthermore, F16BP suppresses depletion of adenosine triphosphate during cold ischemia before intestinal transplantation (Sola et al. 2004). Nevertheless, all aforementioned modified solutions have not been applied into clinical practice as a routine preservation solution. Further investigations are still required to evaluate the usefulness of these elements (Boros et al. 2007).

7.1 Rat Kidney Transplantation

In rat kidney transplantation model, only use of HTK for both perfusion and preservation is inferior to HTK/UW combination or UW solution only, as evidenced by higher histological damage. With respect to long-term preservation, UW is more advantageous over HTK solution. Nevertheless, it is of interest that combination of perfusion with HTK and storage with UW solution can improve graft survival in comparison to UW alone or HTK alone. Perfusion with HTK prior to UW solution preservation can inhibit the production of free radicals, decrease the percentage of initial nonfunctioning kidney, and enhance cellular energy conservation (Schmitz et al. 2006). Nevertheless, UW solution is frequently adopted for experimental studies. Compared with Sacks' solution, use of UW solution can significantly improve graft viability when the cold ischemia time (CIT) is prolonged (24-h) (47 % vs 23 %; $p<0.05$). Furthermore, kidney graft viability will remarkably decrease as cold storage time is prolonged from 24 to 36-h (50 % vs 15 %; $p<0.01$) (Jacobsson et al. 1989). In addition, isotonic solution is superior to hypertonic solution for flushing rat kidney graft, as evidenced by better graft function (serum creatinine level and urea concentration) and morphology after rat renal transplantation (Howden et al. 1983).

7.2 Rat Liver Transplantation

HTK solution can protect liver donor from ischemia and reperfusion injury through decreasing Kupffer cell activation, inhibiting fatty degeneration, intracellular and interstitial edema, and intralobular necrosis in rat liver transplantation (Spiegel et al. 1999). To test the efficacy of HTK, Euro-Collins (EC), and UW preservation solution, rat model of orthotopic syngeneic liver transplantation was performed. The findings revealed that a significantly higher production of bile was observed after utilization of UW preservation (Spiegel et al. 1998). Although additional supplement of buffering agents (bicine, tricine, and histidine) into the "gold standard" UW solution can improve liver metabolic status (total adenylate pool, potential ATP, and energy charge values) during organ retrieval/storage, clinical application of these modified has been observed (Churchill and Kneteman 1998). Apart from storage solution, utilization of novel preservation technique can also result in a significantly improved survival of graft survival. A very recent study reported that supercooling (subzero nonfreezing) preservation and extracorporeal machine perfusion could prolong liver graft preservation for up to 4 days, which was threefold viable preservation during (Berendsen et al. 2014).

7.3 Mouse Heart Transplantation

Indeed, many factors play roles in preservation outcome such as components of solution, perfusion methods, perfusion volume, mouse strain, and storage time. In rat heart transplant model, the endothelium can be injured by means of iron-independent, chloride-mediated impairment, andiron-dependent oxidative damage after cold preservation and reperfusion of organ, resulting in an increased incidence of complication and poorer transplant outcome (Koch et al. 2010). In mouse cardiac transplant model, a maximum of 30-min cold storage and 0.25–0.4 mL normal saline or lactated Ringer's perfusion solution was observed. When organ was kept in cold solution for 1 h through hypothermic machine perfusion with a pump, graft function would fail to recover. Interestingly, mouse strain is closely related with organs' resistance to cold-mediated injury. Heart from C3H (H-2 k) mice had the most resilient ability (Blanchard and Pollak 1985). To optimize the preservation solution, an improved HTK solution including iron chelators (DesfLK) could significantly prolong isogeneic heart (C57BL/6J) graft survival for 24-h cold ischemic preservation through reducing release of creatine kinase and lactate dehydrogenase (HTK: 4.4 days vs. novel: 12.7 days), as evidenced by amelioration of histological injury such as significant lower edema score and inhibition of inflammatory infiltration. It was suggested that chloride-rich solution could improve endothelial cell survival and benefit heart preservation (Wu et al. 2011).

7.4 Mouse Kidney Transplantation

A couple of perfusion and preservation solutions are suggested in mouse kidney transplantation such as normal saline (Han et al. 1999), phosphate buffered saline (Lutz et al. 2007), and lactated Ringer's solution (Wang et al. 2003). The storage solution is normally same with the perfusion solution (Zhang et al. 1995). It is also allowed to use different solution to perfuse donor organ such as UW or HTK solution, which is preferably adopted for prolonged cold ischemia time (Martins 2006). Additional heparin (5 U/mL) may be used in the preservation solution to avoid clot formation.

A couple of studies have been made to investigate the protective effect of various additives of UW solution on transplant kidney during perfusion and storage. Additional use of phosphoenolpyruvate (PEP) (100 mM), a glycolytic intermediate with energy and antioxidative supplementation potentials, can inhibit LDH leakage and histological alterations such as tubulonecrosis and vacuolization during cold preservation of mouse kidney (Ishitsuka et al. 2013). Although prolonged cold ischemia (PCI) can cause autophagy and autophagic flux within cold ischemic kidney, supplementing bafilomycin A1 into UW solution during kidney preservation may significantly reduce apoptotic cell death and autophagy flux (Turkmen et al. 2011).

7.5 Mouse Liver Transplantation

Ice-cold normal saline (Tian et al. 2002), chilled University of Wisconsin (UW) solution (Viaspan solution) (Que et al. 2004), and ice-cold Ringer's solution (Birsner et al. 2004) are used to perfuse liver donor in mouse model. Of them, UW solution at 4 °C is normally recommended to use as it can ameliorate graft from ischemia–reperfusion injury and protect graft from prolonged cold/warm ischemic injury (Que et al. 2004). Additive, such as erythropoietin (EPO), may ameliorate cold ischemic injury and potentiate HTK protective effect on preserved mouse liver through reducing the release of transaminase (Eipel et al. 2012). Use of phosphoenolpyruvate (PEP) (100 mM) is also capable of preventing alteration of oxidative stress parameters of mouse liver during 72-h cold preservation (Ishitsuka et al. 2013). With an increased length of cold preservation, survival after mice syngeneic partial liver transplantation declined from 100, 89, and 0 % at 2 weeks for cold storage (UW solution) of 1, 8, and 12 h, respectively (Wang et al. 2013).

References

Avolio AW, Agnes S, Nure E, Maria G, Barbarino R, Pepe G, Castagneto M. Comparative evaluation of two perfusion solutions for liver preservation and transplantation. Transplant Proc. 2006;38(4):1066–7.

Barlow AD, Hosgood SA, Nicholson ML. Current state of pancreas preservation and implications for DCD pancreas transplantation. Transplantation. 2013;95(12):1419–24.

Belzer FO, Southard JH. Principles of solid-organ preservation by cold storage. Transplantation. 1988;45(4):673–6.

Berendsen TA, Bruinsma BG, Puts CF, Saeidi N, Usta OB, Uygun BE, Izamis ML, Toner M, Yarmush ML, Uygun K. Supercooling enables long-term transplantation survival following 4 days of liver preservation. Nat Med. 2014;20(7):790–3.

Birsner JH, Wan C, Cheng G, Evans ZP, Polito CC, Fiorini RN, Gilbert G, Haines JK, Schmidt MG, Chavin KD. Steatotic liver transplantation in the mouse: a model of primary nonfunction. J Surg Res. 2004;120(1):97–101.

Blanchard JM, Pollak R. Techniques for perfusion and storage of heterotopic heart transplants in mice. Microsurgery. 1985;6(3):169–74.

Boros P, Liu J, Li Y, Bromberg JS. Organ transplantation in rodents: novel applications of long-established methods. Transpl Immunol. 2007;18(1):44–52.

Bretschneider HJ. Myocardial protection. Thorac Cardiovasc Surg. 1980;28(5):295–302.

Churchill TA, Kneteman NM. Investigation of a primary requirement of organ preservation solutions: supplemental buffering agents improve hepatic energy production during cold storage. Transplantation. 1998;65(4):551–9.

Collins GM, Bravo-Shugarman M, Terasaki PI. Kidney preservation for transportation. Initial perfusion and 30 hours' ice storage. Lancet. 1969;2(7632):1219–22.

Dunne JB, Piratvisuth T, Williams R, Tredger JM. Treatment of experimental ischemia/reperfusion injury with S-adenyosylmethionine: evidence that donor pretreatment complements other regimens. Transplantation. 1997;63(4):500–6.

Eipel C, Hubschmann U, Abshagen K, Wagner KF, Menger MD, Vollmar B. Erythropoietin as additive of HTK preservation solution in cold ischemia/reperfusion injury of steatotic livers. J Surg Res. 2012;173(1):171–9.

Guibert EE, Petrenko AY, Balaban CL, Somov AY, Rodriguez JV, Fuller BJ. Organ preservation: current concepts and new strategies for the next decade. Transfus Med Hemother. 2011;38(2): 125–42.

Han WR, Murray-Segal LJ, Mottram PL. Modified technique for kidney transplantation in mice. Microsurgery. 1999;19(6):272–4.

Howden B, Rae D, Jablonski P, Marshall VC, Tange J. Studies of renal preservation using a rat kidney transplant model. Evaluation of citrate flushing. Transplantation. 1983;35(4):311–4.

Ishitsuka Y, Fukumoto Y, Kondo Y, Irikura M, Kadowaki D, Narita Y, Hirata S, Moriuchi H, Maruyama T, Hamasaki N, Irie T. Comparative effects of phosphoenolpyruvate, a glycolytic intermediate, as an organ preservation agent with glucose and N-acetylcysteine against organ damage during cold storage of mouse liver and kidney. ISRN Pharmacol. 2013;2013:375825.

Jacobsson J, Wahlberg J, Tufveson G. Influence of different preservation solutions and intentional hemodilution of recipient on viability of preserved and transplanted rat kidneys. Transpl Int. 1989;2(2):117–20.

Koch A, Loganathan S, Radovits T, Sack FU, Karck M, Szabo GB. Deferoxamine, the newly developed iron chelator LK-614 and N-alpha-acetyl-histidine in myocardial protection. Interact Cardiovasc Thorac Surg. 2010;10(2):181–4.

Lee CY, Mangino MJ. Preservation methods for kidney and liver. Organogenesis. 2009;5(3): 105–12.

Lutz J, Lu R, Strobl M, Huang H, Deng M, Wang M, Ouyang N, Heemann U. ICOS/B7RP-1 interference in mouse kidney transplantation. Transplantation. 2007;84(2):223–30.

Martins PN. Technique of kidney transplantation in mice with anti-reflux urinary reconstruction. Int Braz J Urol. 2006;32(6):713–8; discussion 719–720.

Natori S, Higuchi H, Contreras P, Gores GJ. The caspase inhibitor IDN-6556 prevents caspase activation and apoptosis in sinusoidal endothelial cells during liver preservation injury. Liver Transpl. 2003;9(3):278–84.

Que X, Debonera F, Xie J, Furth EE, Aldeguer X, Gelman AE, Olthoff KM. Pattern of ischemia reperfusion injury in a mouse orthotopic liver transplant model. J Surg Res. 2004;116(2): 262–8.

Schmiedt CW, Schwab MC, Dubielzig RR, Murphy CJ, McAnulty JF. Trophic factor supplemented UW solution reduces intimal hyperplasia in the rat aortic transplant model. Cryobiology. 2007;54(2):204–11.

Schmitz V, Klawitter J, Bendrick-Peart J, Haschke M, Beckey VE, Laudi S, Neumann U, Schoening W, Neuhaus P, Christians U, Puhl G. Impact of organ preservation using HTK for graft flush and subsequent storage in UW in rat kidney transplantation. Eur Surg Res. 2006;38(4):388–98.

Sola A, De Oca J, Alfaro V, Xaus C, Jaurrieta E, Hotter G. Protective effects of exogenous fructose-1,6-biphosphate during small bowel transplantation in rats. Surgery. 2004;135(5): 518–26.

Southard JH, Belzer FO. Organ preservation. Annu Rev Med. 1995;46:235–47.

Spiegel HU, Schleimer K, Kranz D, Oldhafer KJ. Organ preservation with EC, HTK, and UW solutions in orthotopic liver transplantation in syngeneic rats. Part I: functional parameters. J Invest Surg. 1998;11(1):49–56.

Spiegel HU, Schleimer K, Freise H, Diller R, Drews G, Kranz D. Organ preservation with EC, HTK, and UW, solution in orthotopic rat liver transplantation. Part II. Morphological study. J Invest Surg. 1999;12(4):195–203.

Tian Y, Rudiger HA, Jochum W, Clavien PA. Comparison of arterialized and nonarterialized orthotopic liver transplantation in mice: prowess or relevant model? Transplantation. 2002;74(9):1242–6.

Turkmen K, Martin J, Akcay A, Nguyen Q, Ravichandran K, Faubel S, Pacic A, Ljubanovic D, Edelstein CL, Jani A. Apoptosis and autophagy in cold preservation ischemia. Transplantation. 2011;91(11):1192–7.

Wang M, Bai J, Baumann M, Heemann U. New model for simultaneous heart and kidney transplantation in mice. Microsurgery. 2003;23(2):164–8.

Wang G, Hu B, Li Z. Cold ischemia/reperfusion injury in a mouse model of partial liver transplantation. J Surg Res. 2013;181(2):337–41.

Wu K, Turk TR, Rauen U, Su S, Feldkamp T, de Groot H, Wiswedel I, Baba HA, Kribben A, Witzke O. Prolonged cold storage using a new histidine-tryptophan-ketoglutarate-based preservation solution in isogeneic cardiac mouse grafts. Eur Heart J. 2011;32(4):509–16.

Zhang Z, Schlachta C, Duff J, Stiller C, Grant D, Zhong R. Improved techniques for kidney transplantation in mice. Microsurgery. 1995;16(2):103–9.

Chapter 8
Rat Kidney Transplantation

Weihua Gong

Rat kidney transplantation is an important experimental model to mimick clinical scenarios in view of the challenging techniques of mouse kidney transplant model and a similarity between rat and human physiology. Indeed, the first heterotopic/orthotopic kidney transplantation in rat was respectively established by Fisher B and Lee S in 1965 (Fisher and Lee 1965) and by Daniller A et al. in 1968 (Daniller et al. 1968). Thereafter, this model has been widely used for experimental studies and enhanced our understanding of transplant immunology, the rejection process, and prevention strategies (Gong et al. 2009). To achieve the techniques better and faster, many attempts have been made to modify surgical procedures and improve this model including the sleeve anastomosis, a nonsplinted ureteroureterostomy, utilization of the donor inferior vena cava, and ureter implantation technique. Herein, we introduce a reliable, reproducible transplant model with fewer complications on basis of the successful experimental practice in Germany. Although utilization of both donor kidneys is practically feasible, short length of right renal vessels makes transplantation technically more difficult. Therefore, this transplant model is established using left kidney.

8.1 Rat Kidney Transplant Models

The most commonly used acute rejection model for kidney transplantation is the rat Dark Agouti (DA) to Lewis (LEW), whereas the chronic rejection model is the Fischer 334 (F344) to LEW. These inbred strains of the former combination are

W. Gong, MD, PhD
Department of Surgery, The Second Affiliated Hospital of Zhejiang
University School of Medicine, Hangzhou, People's Republic of China
e-mail: weihua.gong@gmail.com

W. Gong (ed.), *Rodent Transplant Medicine*,
DOI 10.1007/978-94-017-9472-5_8

MHC fully mismatched and the latter rats are haploidentical and different only at minor histocompatibility (non-MHC) loci.

8.2 Animals

Naïve male 200–250 g inbred rats are normally used for the experiments. Lewis rats (LEW; $RT1^l$) and Dark Agouti (DA; $RT1^a$) animals can be used as fully MHC-mismatched donors and recipients for allogeneic transplant studies. The rats are kept under standard conditions with free access to rat food and water ad libitum. All animal experiments are performed under the permission of the local authorities and according to the German Animal Protection Acts.

8.3 Anesthesia

Anesthesia is induced and maintained with continuous inhalation of O_2 and isoflurane (Forane™, Abbot, Germany) during whole transplantation procedures.

8.4 Donor Surgery

The anesthetized donor DA rat is fixed in a supine position with paper tapes. The hair of surgical area is shaved and scrubbed by using Betadine, 70 % alcohol solution for disinfection. Somatic heparinization is achieved by injection with 0.6 ml heparinized saline (4,000 U/ml) via the dorsal penile vein. A long midline incision is made from the xiphoid to the pubic symphysis. The visceral organs are mobilized to the right side of the abdominal cavity. The left ureter is located and freed from the retroperitoneum to the level of iliolumbar vein (Fig. 8.1d). Care should be taken to protect the periureteric tissue, especially the small periureteral nourishing vessels. The left renal vein (Fig. 8.1b) and artery (Fig. 8.1a) are bluntly dissected and separated from each other. Herein, careful maneuver should be done to avoid impair renal arterial intima; otherwise, thrombosis is likely to form after transplant surgery. To avoid vasospasms especially for small renal artery, 1 % lidocaine or xylocaine may be applied around the renal pedicle. Subsequently, the adventitia of kidney (Fig. 8.1c) is well maintained and fat on its surface can be almost removed. The inferior vena cava (IVC) (Fig. 8.1e) and abdominal aorta (Fig. 8.1f) are clamped beneath the diaphragm by using the hemostatic forceps, and inferior vena cava is transected below the level of the left renal vein. The renal allograft is immediately perfused by using 20 ml precooling normal saline (4 °C) with a 20-gauge catheter (Venflon™, BD, Germany) and 10 ml University of Wisconsin (UW) solution (4 °C) (Dupont Pharma, USA) with 10-gauge catheter (Venflon™, BD, Germany), which are inserted through the aorta bifurcation until the kidney becomes uniformly pale

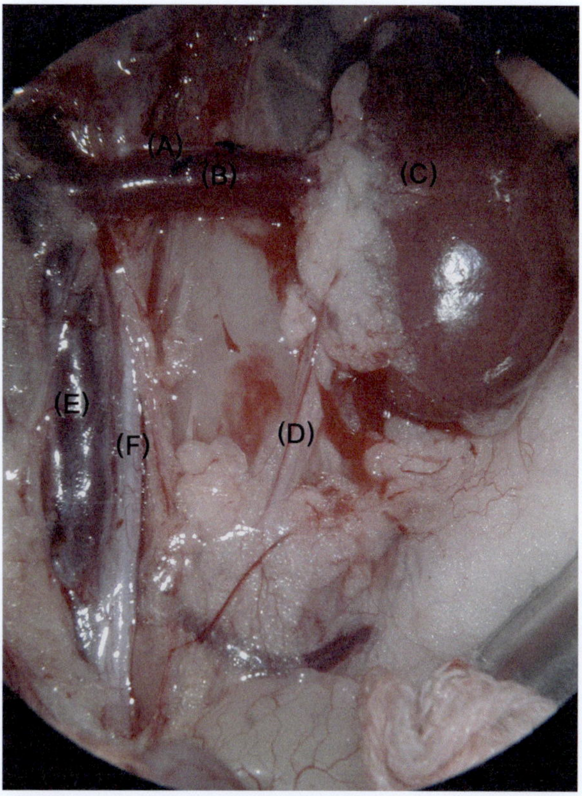

Fig. 8.1 During donor procurement, graft including left renal artery (*A*), renal vein (*B*), kidney (*C*), ureter (*D*), inferior vena cava (*E*), and abdominal aorta (*F*) is carefully dissected

and the perfusate is clear (Fig. 8.2). The ureter is incised at the level where it intersects with the iliolumbar vein. The procured donor kidney with partial attached ureter is stored in UW solution at 4 °C for 30–60 min.

8.5 Recipient Surgery

After the rat is anesthetized, the hair of surgical area is shaved and placed supine with its tail toward microsurgeon. The skin is disinfected using Betadine, 70 % alcohol solution. When the alcohol appears dry, the draping process begins. After a midline laparotomy is made, the recipient's abdominal cavity is exposed using self-retaining retractors. The gut is reflected to the left side of the abdominal cavity to expose the infrarenal IVC and aorta. The left renal vessels are prepared as described above. The ureter is transected at the level where it intersects with iliolumbar vein (Fig. 8.3c). Given the appearance of two renal arteries, the smaller one is ligated with 7–0 silk suture. The renal vein and artery are respectively clamped with vascular clips next to abdominal aorta and IVC. Both renal vessels are incised as near as possible to the left kidney (Fig. 8.3a, b). The free ends of these two vessels are irrigated with normal saline to remove intraluminal remaining blood. The native

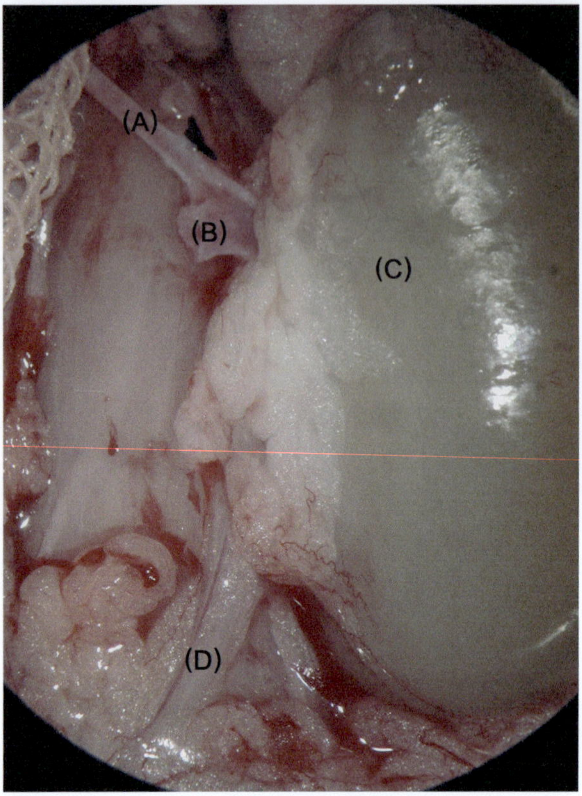

Fig. 8.2 Donor graft (*C*) is perfused using precooling normal saline until the kidney becomes uniformly pale. Renal vein (*B*) is transected close to renal hilum, whereas renal artery (*A*) is elliptically incised on the wall of abdominal aorta (Courtesy of Dr. Christian Denecke from Medizinische Universitaet Innsbruck, Austria)

kidney is replaced with the donor graft. The contralateral kidney can be removed at 3 or 7 or 10 posttransplant. The different timepoints for the nephrectomy have been reported including the same day with kidney transplant surgery.

8.6 Vascular Anastomosis

The graft is mobilized from storage solution and positioned in the left side of the recipient's abdominal cavity. The graft (Fig. 8.4c) is surrounded with ice-cold moisture pad during vascular anastomosis. Subsequently, the donor's renal artery and vein are respectively anastomosed end to end to the recipient's renal artery (Fig. 8.4a) and vein (Fig. 8.4b). First, two stay sutures are carried out with 10/0 prolene (Ethicon, USA) in the opposed ends of the arteries at around 180° to each other. On each side 3–4 stitches are interruptedly applied. To clearly expose the posterior wall of the renal artery, vessel clips are reversed up and down. With respect to venous

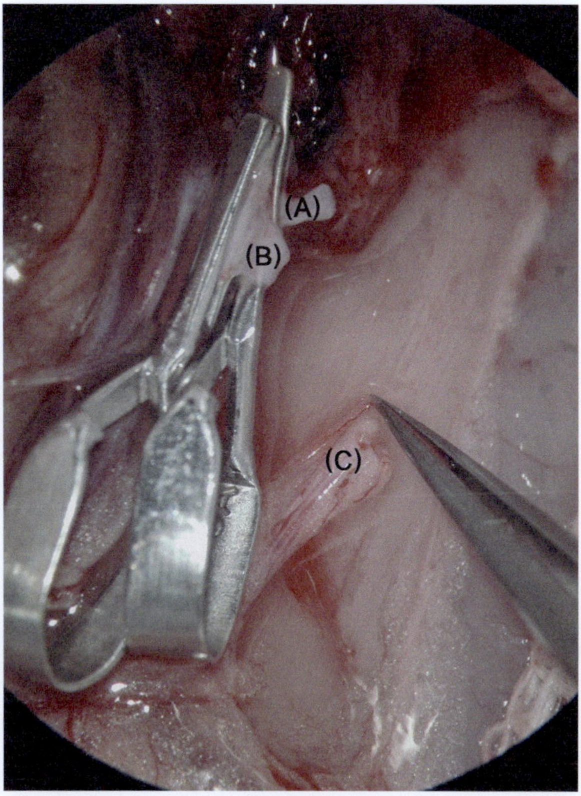

Fig. 8.3 Recipient's left kidney is removed and the ends of renal vessels (*A*, *B*) and ureter (*C*) are shown

anastomosis, two stay stitches are performed with 10/0 prolene in the opposed ends of the veins at around 150° to each other, leaving a relatively long tail for each stay stitch, by means of which renal veins can be conveniently held. The backside of the vein is firstly sutured. The whole circumference of renal vein anastomosis is continuously stitched in a clockwise direction. At the end, the edges of the veins are gently stretched to avoid anastomotic construction.

8.7 Reperfusion

As the renal vascular reconstruction is done, the clip for the renal vein should be firstly released. Blood returns to the kidney graft in dark red (Fig. 8.5). Subsequently, the clip for the renal artery is immediately released. Given that a little bleeding occurs, the anastomotic site can be gently pressed by using cotton swab for a few seconds. Renal artery is reclamped and additional stitches are necessary if bleeding persists.

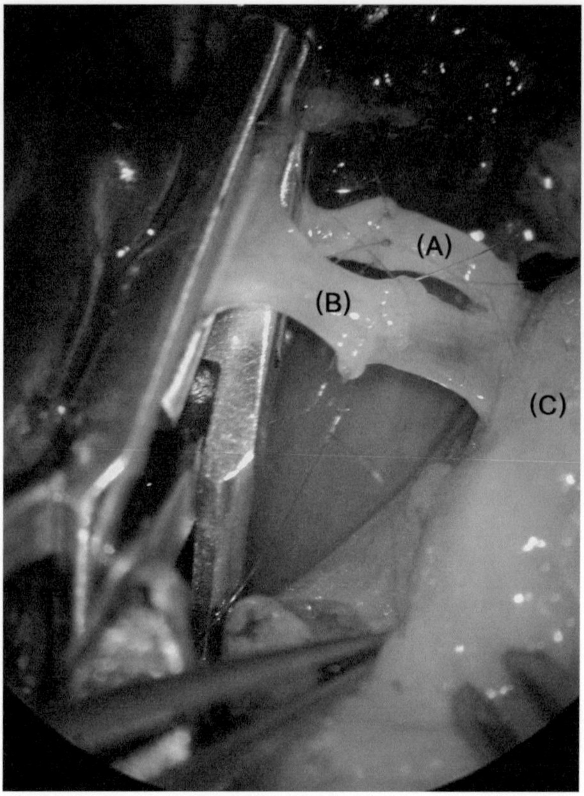

Fig. 8.4 After kidney graft (*C*) is positioned in situ, renal artery (*A*) and vein (*B*) are end to end anastomosed by interrupted and continuous suturing respectively

8.8 Ureter Anastomosis

Care should be taken not to twist any ureter before ureter reconstruction. The ureteral end-to-end reconstruction is done using 10/0 prolene with 4–6 interrupted stitches (Fig. 8.6). Given the appearance of obvious bleeding from the small periureteral nourishing vessels, one or two stitches can be applied to ligate the ends of the vessels. Based on autopsy, the recipients with postoperative complications, such as urine leakage and hydronephrosis, are excluded from research analysis. Indeed, after graft is reperfused, blood supply to the donor's ureter is restored and consequently urine is produced. Importantly, close observation of urine passage through anastomotic site and undisturbed peristaltic waves can be utilized for a prima facie ureteric anastomotic patency after ureter reconstruction is done (Fig. 8.7).

Before the abdomen is closed, the "free" kidney graft is fixed by suturing the adventitia of graft to the retroperitoneum. The two layers of abdomen are continuously stitched using 3–0 or 4–0 absorbable sutures.

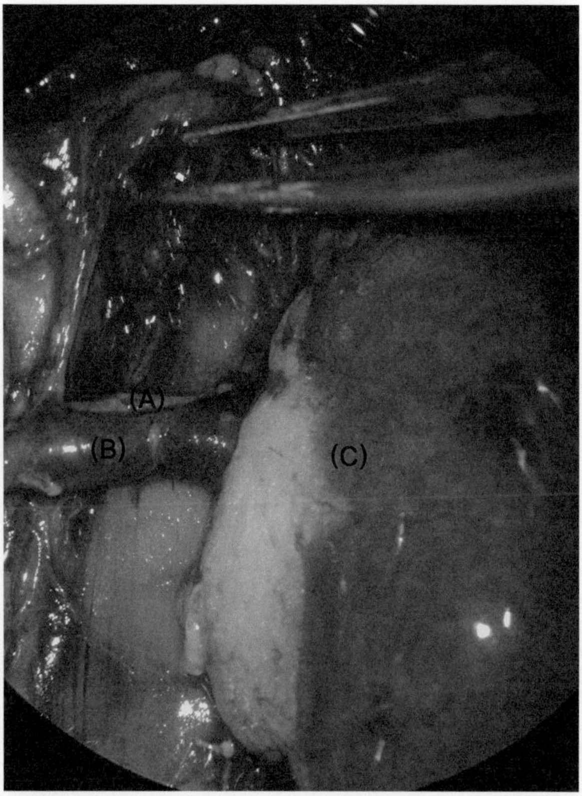

Fig. 8.5 After microclips are released, kidney graft (*C*) is reperfused and turns dark red. No constriction is observed for renal arterial (*A*) and venous (*B*) anastomosis

8.9 Postoperative Care

After the abdomen is closed, the Lewis rat is routinely given 50 mg Novaminsulfon (Ratiopharm 1 GmbH, Ulm) as an analgesic. Infrared light is applied to warm up the rat until consciousness. After transplant surgery, rat is returned to the cage with free access to food and water at all times.

8.10 Graft Assessment

The graft's function is normally determined by creatinine clearance and proteinuria. 24-h urine output is collected by placing the recipient in a metabolic cage with free access to food and water (Harvard Apparatus, Holliston, MA). Proteinuria can sensitively indicate the chronic alteration of graft function. Protein excretion (mg/24 h) is measured by precipitation with 20 % CCl3COOH. Turbidity is determined at a

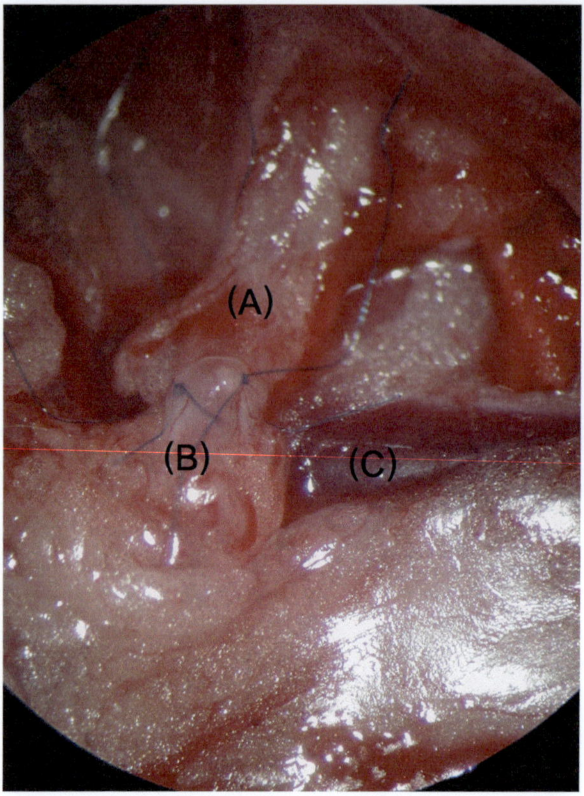

Fig. 8.6 Donor's ureter (*A*) is end to end anastomosed to recipient's ureter (*B*) at the level of recipient's iliolumbar vein (*C*)

wavelength of 415 nm by using a Hitachi 911 analyzer (MYCO Instrumentation Source, Inc., Washington, USA). Plasma creatinine level is measured by portable equipment (Roche Reflotron Plus, Roche Diagnostics).

Blind histological analysis is done to evaluate glomerulosclerosis of kidney graft based on Banff mesangial matrix score. The presence of glomerular fibrosis stained by the Periodic Acid Schiff (PAS) is defined as: grade 1 = 0–25 %; grade 2 = 25–50 %; grade 3 = 50–75 %; and grade 4 = 75–100 % sclerosis of glomerulus. The glomerulosclerosis index = (grade1 × glomeruli amount + grade2 × glomeruli amount + grade3 × glomeruli amount + grade 4 × glomeruli amount)/total glomeruli amount.

8.11 Discussion

Currently, with the fast advance in microsurgical techniques, a high long-term survival rate of rat kidney transplantation can be routinely achieved with a high rate of vascular patency (up to 95 %) (Gu et al. 2002). Another reason for satisfaction

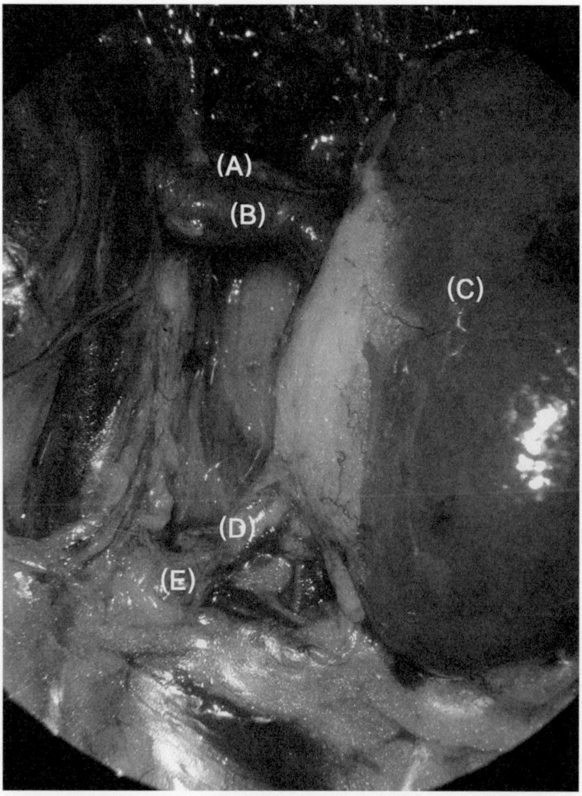

Fig. 8.7 Holistic view after kidney transplant surgery. (*A*) Renal artery; (*B*) Renal vein; (*C*) Transplanted kidney; (*D*) Donor's ureter; and (*E*) Recipient's ureter

success rate is the bigger size of rat compared with mouse, which can cause a profound difference of microsurgical difficulties. The utilization of rat for research studies is available for covering most aspects of transplant biology (Oesterwitz et al. 1983).

Although various modifications of the surgical techniques were made by researchers, postoperative complications still can be observed in practice. With respect to vascular reconstruction, the procurement of donor organ is important. The cut-end of donor's renal vein is achieved close to the inferior vena cava, ensuring sufficient size of the diameter of renal vein.

Indeed, urinary reconstruction is a critical point for a success of rat kidney transplantation, i.e., the restoration of the continuity of the urinary tract represents one of the major problems in rat renal transplantation. It was found that technique of end-to-end ureteric anastomosis is superior to others based on lower frequency of postoperative complications and operation time (Oesterwitz et al. 1983). On the other hand, end-to-end ureterostomy is the most physiologically effective technique. An important issue from the viewpoint of surgical techniques is a strict adherence to essential principles of microsurgery for such small luminal structures of ureter. Too many sutures will result into an obvious ureter constriction at anastomotic site, whereas too fewer knots may cause urine leakage in the rat abdomen. The cut-end of ureter may be trimmed into oblique shape when necessary. The occurrence of stenosis of ureteric anastomosis should be completely avoided. Preservation of periureteric tissue is necessary, which will keep the ureter's blood supply. The cut-end of periureteral nourishing vessels can be ligated or carefully cauterized according to our own experience in order to avoid hemorrhage of the urethra (Ge and Gong 2011).

The first chronic rejection model in rat was reported by Dr. White E et al. in 1969, in which rat Fischer 334 (F344) and Lewis (LEW) were respectively used as donor and recipient (White et al. 1969). In this combination, only minor histocompatibility (non-MHC) loci were different between those two inbred haploidentical strains. In this rat kidney transplant model, slow impairment of kidney allograft is observed, as evidenced by parenchymal alteration, nephron loss, and renal dysfunction after 48 weeks (Marco 2006). Differently, the most commonly used acute rejection model for kidney transplantation is rat DA to Lewis model, in which rejection episode normally occurs at day 5–7. Massive macrophage and T lymphocytes are infiltrated into allografts, leading to damage of graft architecture and allograft failure (Kakuta et al. 2012).

References

Daniller A, Buchholz R, Chase RA. Renal transplantation in rats with the use of microsurgical techniques: a new method. Surgery. 1968;63(6):956–61.

Fisher B, Lee S. Microvascular surgical techniques in research, with special reference to renal transplantation in the rat. Surgery. 1965;58:904–14.

Ge F, Gong W. Strategies for successfully establishing a kidney transplant in a mouse model. Exp Clin Transplant. 2011;9(5):287–94.

Gong W, Klopfel M, Reutzel-Selke A, Jurisch A, Vogt K, Haase S, Hoflich C, Polenz D, Gerstmayer B, Tomiuk S, Volk HD, Pascher A, Sawitzki B. High weight differences between donor and recipient affect early kidney graft function–a role for enhanced IL-6 signaling. Am J Transplant. 2009;9(8):1742–51.

Gu YL, Dahmen U, Dirsch O, Broelsch CE. Improved renal transplantation in the rat with a nonsplinted ureteroureterostomy. Microsurgery. 2002;22(5):204–10.

Kakuta Y, Okumi M, Miyagawa S, Tsutahara K, Abe T, Yazawa K, Matsunami K, Otsuka H, Takahara S, Nonomura N. Blocking of CCR5 and CXCR3 suppresses the infiltration of macrophages in acute renal allograft rejection. Transplantation. 2012;93(1):24–31.

Marco ML. The Fischer-Lewis model of chronic allograft rejection–a summary. Nephrol Dial Transplant. 2006;21(11):3082–6.

Oesterwitz H, Althaus P, May G, Schroder K, Strobelt V, Kaden J. Microvascular surgery in experimental and clinical kidney transplantation. Int Urol Nephrol. 1983;15(4):347–57.

White E, Hildemann WH, Mullen Y. Chronic kidney allograft reactions in rats. Transplantation. 1969;8(5):602–17.

Chapter 9
Rat Orthotopic Liver Transplantation

Jinzhen Cai and Junwu Cai

Rat orthotopic liver transplantation (OLT) has been generally accepted as a studies model in the research of organ preservation, transplant pathology, physiology, and immunology since it was firstly developed by Dr. Sun Lee et al. (1973). In 1979, Kamada and Calne reported "two-cuff technique" for portal vein and infrahepatic vena cava (Kamada and Calne 1979), which significantly shortened anhepatic phase to 15–20 min and increased 1-week survival rates to 70 %. Afterward, their successful techniques have become widely used in practice (Hori et al. 2010). Herein, we present the detailed surgical procedures for a rat model of OLT with "two-cuff technique" and without hepatic arterial reconstruction.

9.1 Animal

Male Lewis rats weighing between 230 and 250 g are used as the donor and recipient for the liver transplantation model. The rats are housed in clean condition and free access to a standard diet and water ad libitum before transplant surgery.

9.2 Anesthetization

Anesthesia is induced by inhalation of 4 vol% isoflurane in 100 % oxygen at a flow rate of 4 L/min, and 1.5 vol% at 2 L/min for the maintenance. Buprenorphine (0.1 mg/kg) is subcutaneously injected as an analgesic after transplant surgery.

J. Cai, MD, PhD (✉) • J. Cai
Department of Transplantation, Transplant Center, Tianjin First Central Hospital, Tianjin Medical University, Tianjin, People's Republic of China
e-mail: caijinzhen@sina.com

W. Gong (ed.), *Rodent Transplant Medicine*,
DOI 10.1007/978-94-017-9472-5_9

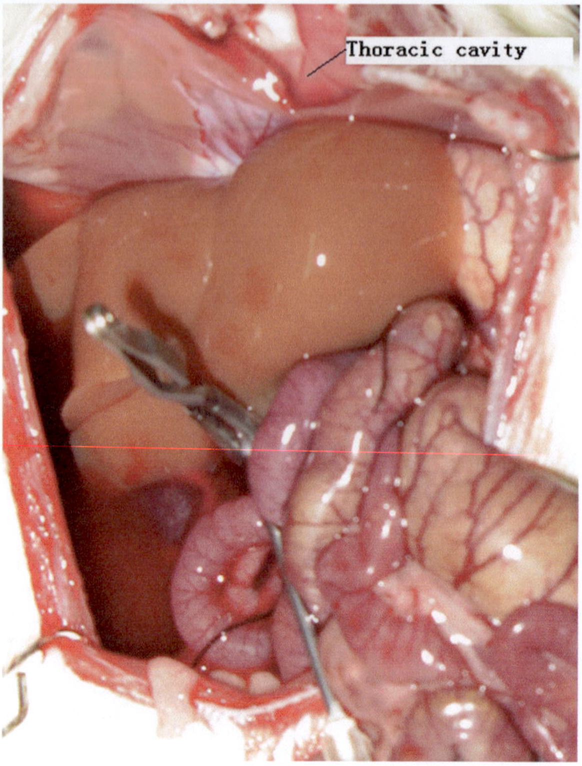

Fig. 9.1 Donor perfusion with solution and then harvested. The donor procedure requires approximately 30–45 min. The donor liver is flushed by preservation solution and subsequently immediately harvested

9.3 Donor Procedure

The rat is positioned on a pad and the four arms are fixed. We shave the fur of abdominal area and sterilize the skin. The midline incision is made to open the abdomen. The xiphoid is pulled toward the head by using a mosquito forceps and clamp. The subcostal retractors are used to open the surgical field (Fig. 9.1). The falciform ligament and left triangular ligament are dissected. The left phrenic vein is divided. The intestines are moved onto the left side of the rat abdomen and covered with wet gauze. The IHVC is isolated from the retroperitoneal tissue. The right adrenal vein is ligated. A stent is inserted into the bile duct and secured with a 6-0 silk thread. The portal vein is divided from the pyloric and splenic veins. The gastroduodenal artery is dissected and ligated. 500 IU of Heparin-Natrium in 2 ml of normal saline solution is injected through the penile vein. Three minutes later, the IHVC is clamped close to the right renal vein with a mosquito forceps. The portal vein is clamped with a disposable microvessel clamp below the stump of the splenic vein. The anterior wall of the portal vein is incised. An 18-gauge catheter is inserted into the portal vein. The liver is perfused in situ with 60 ml of cold histidine-tryptophan-ketoglutarate (HTK) solution at a hydrostatic pressure of 20 cm H_2O. Afterward, the diaphragm is opened immediately and the intrathoracic vena cava is transected. The anterior wall of the IHVC is opened for the release of liver donor perfusate. The liver is excised by dissecting the IHVC between the liver and the right renal vein. The portal vein is dissected below the stump of the splenic vein. The diaphragm, the remaining ligaments at the back of the liver, and the right adrenal vein are subsequently dissected. The donor liver is stored in cold preservation solution in a metal cup, which is mounted in a plastic box full of crushed ice.

9.4 Donor Back Table Procedure

The liver graft is kept in the metal cup full of ice-cold HTK solution. The procedure of back table needs approximately 30 min, by which "two-cuff" is accomplished (Fig. 9.2). A Bulldog clip in a bridging position over the cup is used to clamp the portal venous trunk. The portal vein is pulled through the cuff and clamped with the

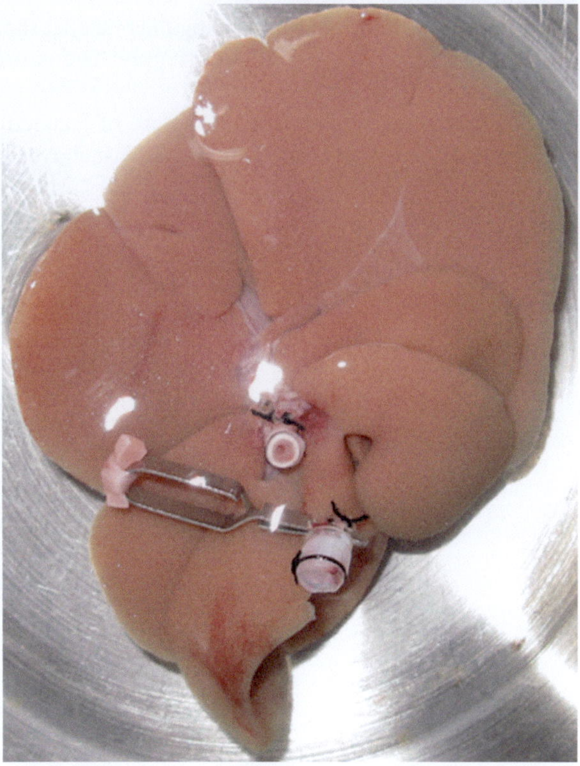

Fig. 9.2 The "two-cuff" attachment of portal vein and IHVC (the infrahepatic vena cava)

extension of the cuff at the 12 o'clock position. A 6-0 silk thread is used to secure the portal vein. With regard to the SHVC (the suprahepatic vena cava), the same method is adopted to attach the cuff into the SHVC. The liver graft is preserved at 4 °C in HTK solution in a cold-water bath.

9.5 Recipient Procedure

The rat recipient procedure of the graft implantation is shown in Figs. 9.3, 9.4, and 9.5. It requires approximate 45 min, which includes nearly 20 min of anhepatic time. The midline incision is made to open abdomen for the donor operation. A wet swab is used to move over the right side of the duodenum and move out the whole intestines to better expose a surgical field around the IHVC (infrahepatic vena cava). The left lateral and median lobes are moved into the left subphrenic cavity. The right lateral lobe is retracted upward with a wet swab. The IHVC is isolated from the retroperitoneal tissue. The right adrenal vein is ligated and divided. The surgical board is rotated, which renders the liver direct toward the left side. The ligament is dissected around the back of the liver. The right lateral lobe is repositioned to the anatomical position. The bile duct is transected just below the branch from the caudate lobe. The soft tissues surrounding the bile duct should be preserved as much as possible to ensure nourishing blood supply. The gastroduodenal artery is divided and ligated.

After the intravenous injection of 2 ml of lactated Ringer solution, the IHVC with a metal microvessel clamp is clamped just above the right renal vein. The portal

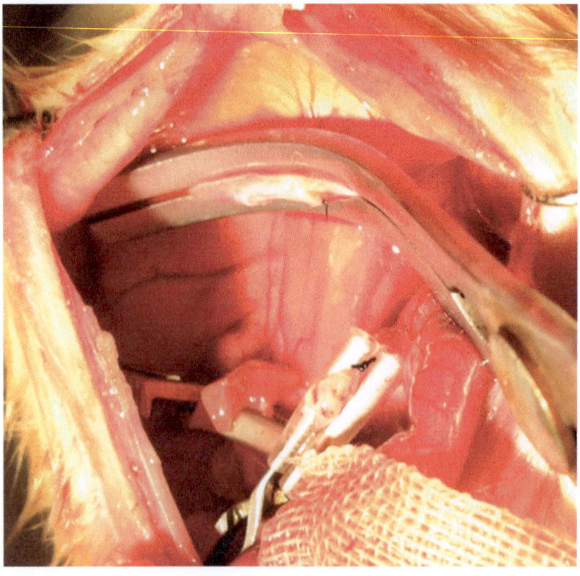

Fig. 9.3 The liver is resection at the anhepatic phase

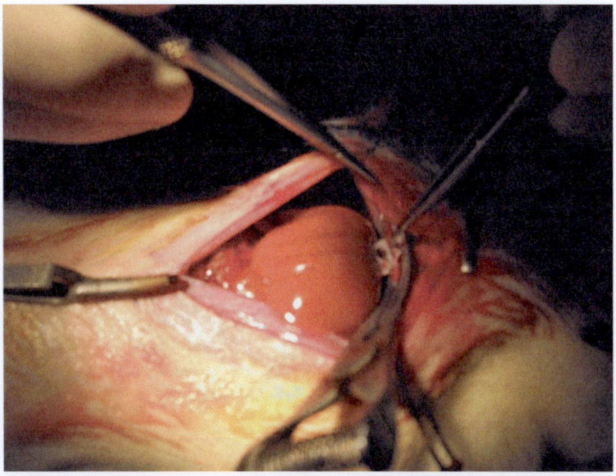

Fig. 9.4 The donor liver is orthotopically positioned. The anastomosis of the SHVC (the suprahepatic vena cava) is accomplished by running suture

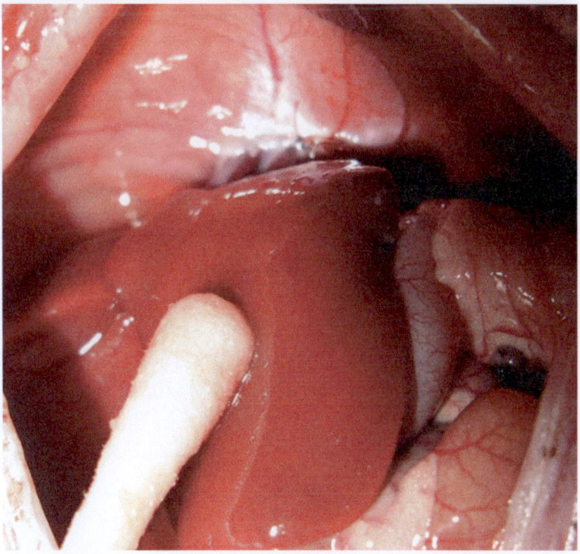

Fig. 9.5 The reperfusion of the donor liver

vein is clamped at the level of its bifurcation in the liver hilum by a mosquito forceps from the left side of the rat. The SHVC together with the diaphragm is clamped from the right side by a peripheral vascular clip. The recipient native liver is excised by dissecting the SHVC, the portal vein, and the IHVC at the following levels: the SHVC, at the border between the SHVC and the liver; and the portal vein, and the IHVC, slightly below the middle point between the liver and the right renal vein.

The liver graft is orthotopically positioned. The SHVC is anastomosed by running suture. The recipient portal vein is clamped at its confluence with the pyloric vein by using a disposable micro vessel clamp from the right. The mosquito forceps is fixed for better clamping the portal vein in the clay, and pulling the tip of

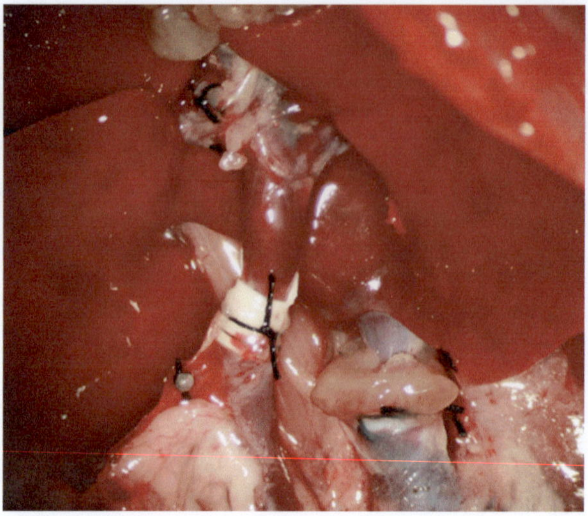

Fig. 9.6 The anastomosis portal vein and IHVC: insert the cuff into the recipient portal vein deeply, and secure it with a circumferential 6-0 silk thread. Use the same way to anastomosis IHVC

the forceps toward the liver hilum. The anterior wall of the portal vein is incised just below the jaw of the mosquito forceps. The inside of the recipient portal vein and the cuff is washed by using lactated Ringer solution. The anterior wall of the incision is held with a straight micro forceps in the left hand and an extension of the cuff with a curved microforceps in the right hand. The cuff is inserted into the recipient portal vein deeply and secured with a circumferential 6-0 silk thread. The same way is taken to reconstruct IHVC (Fig. 9.6). The clamps for the portal vein, IHVC, and SHVC are released in order. Subsequently, the liver is reperfused.

As the clamps on portal vein and SHVC are released, the graft liver regains blood supply and the anhepatic phase is completed. The bile duct is connected end to end via Teflon catheter and fixed with 5-0 silk suture. No attempt is made to reestablish the blood flow of the hepatic arterial. In the end, the abdominal incision is closed with a continuous 3-0 suture.

9.6 Discussion

The needle and thread are normally used for vascular anastomosis. However, suturing might cause several detrimental aspects. The penetrating needle induces vascular wall damage, which influences the healing response. As nonabsorbable suture material is left, this intraluminal foreign body may result in an inflammatory reaction, thrombocyte aggregation, impaired endothelial function, intimal hyperplasia, and subsequent stenosis (Zeebregts et al. 2003). Moreover, performing sutured vascular anastomoses demands extensive training for a high patency rate (Werker and Kon 1997), especially for relatively difficult OLT model in rat.

The “two-cuff technique” in the rat liver transplantation model is widely use in the research of liver transplantation. The techniques for portal vein and infrahepatic

IVC described by Kamada and Calne have successfully improved long-term survival rates up to 90 % (Kamada and Calne 1979). Although "three-cuff model" for the three venous anastomoses was introduced later on by Miyata in 1980, the diameter of the intrathoracic segment of the vena cava was too thin to ensure the blood flow. Furthermore, there was a risk of kinking and torsion of vessels. These have hampered the wide spread of this method. Tsuchimoto et al. (1988) and Yasushi et al. (1992) also described cuff technique for SHVC anastomoses. An important issue is the difficulty in inserting the cuffed donors' SHVC into the recipients' SHVC. In addition, making the ligation becomes difficult owing to the short anterior wall of the recipients' SHVC. The above-mentioned pitfalls have also been described by other studies using cuff technique for SHVC anastomosis (Tan et al. 2005; Oldani et al. 2008). Our transplant model presents the reconstruction procedures for each vessel and the bile duct which are performed by the following methods: a 7-0 polypropylene continuous suture for the supra- and infrahepatic vena cava; a cuff technique for the portal vein; and a stent technique for the bile duct.

9.7 Selection for Recipient and Donor

Long time of donor fasting before operation will cause acute hepatic steatosis, and no fasting will affect the quality of the surgical procedure. As per our experience, the ideal time of donor and recipient preoperative fasting is suggested as 6 h.

9.8 Narcotic Control and Choice

Most of the current selections are ether, ketamine, and other single or combined anesthesia; ether anesthesia has the effect of opening a stable, nontoxic to the liver, postoperative awake, etc., but will increase respiratory secretions which must advance the intramuscular injection of atropine, inhaled volume accuracy poor. Chloral hydrate (300 mg/kg) sedation strong analgesic effect, intraperitoneal administration to 2 min after surgery, can be maintained for about 1 h, soon after waking up. No case will die due to anesthesia.

9.9 Donor Liver Surgery and Back Table Procedure

Many researchers advocate the grand cross of the abdominal incision. However, cross-incision will cause excessive bleeding. If abdominal exposure is too large, liver damage or even liver shock will occur. Therefore, we use vertical incision with pull hook corners retractor, which is simple with completely full exposure. Before perfusion, we dissect the ligament of the liver, and IVC. It would reduce the exposure of the liver and excess liver injury can be avoided. Furthermore, it will also decrease the residence time of the cold perfusion liver in the abdominal cavity. The heparin infusion before whole liver is the key point of the donor quality. In terms of the short half-life of heparin, timing is important. We give the heparin just before the portal perfusion.

Portal vein is pulled through the cuff, and ligation splenic vein with the 7/0 wire, using saline to check the patency of the portal vein. Vena cava cannula can be dug out of the renal vein. Liver IVC reserved diaphragm ring, diameter without excessive expansion, hung left corner to the left phrenic vein marks.

9.10 Receptors Surgery

Lumbar back pad bridge, two pull hook to the head can be stretched into a triangle which could fully reveal the liver and IVC, as well as reduce trauma. Ligate the left phrenic vein properly away from the liver to avoid bleeding after reperfusion. Keep enough space for the back wall of IVC. The posterior wall of hepatic vein located deeply in the diaphragm is easy to neglect. Bleeding will be caused if the vessel is injured, so ligation is necessary.

During the anhepatic phase, early trainer should not blindly seek fast. It is likely to cause bleeding and anastomotic stricture after perfusion. The time within 26 min does not directly lead to surgical failure. Portal vein and bile duct anastomosis is the most difficult step of the operation for a single operator. By taking the left corner and the front wall line, the surgeon forceps jaws right corner. A good open lumen can be exposed by using this method, which will reduce the difficulty.

The main reasons for transplantation failure are anastomotic stenosis and intraoperative bleeding. Anastomotic stricture is frequently caused by primary technique, anastomotic thrombosis, etc. Rat blood volume is less, and the tolerance will become poorer for blood loss. The common bleeding location is in the liver IVC corner and the back wall. The bleeding must be completely hemostasis before the abdomen is closed.

In short, each step from donor procurement to recipient implantation has to be carefully performed for successful rat liver transplantation. A 90 % of survival rate may be achieved after a 3-month microsurgical training.

References

Harihara Y, Sanjo K, Idezuki Y. A modified cuff technique for suprahepatic vena cava anastomosis in rat liver transplantation. Transplantation. 1992;53(3):707–9.

Hori T, Nguyen JH, Zhao X, Ogura Y, Hata T, Yagi S, Chen F, Baine AM, Ohashi N, Eckman CB, Herdt AR, Egawa H, Takada Y, Oike F, Sakamoto S, Kasahara M, Ogawa K, Hata K, Iida T, Yonekawa Y, Sibulesky L, Kuribayashi K, Kato T, Saito K, Wang L, Torii M, Sahara N, Kamo

N, Sahara T, Yasutomi M, Uemoto S. Comprehensive and innovative techniques for liver transplantation in rats: a surgical guide. World J Gastroenterol. 2010;16(25):3120–32.

Kamada N, Calne RY. Orthotopic liver transplantation in the rat. Technique using cuff for portal vein anastomosis and biliary drainage. Transplantation. 1979;28(1):47–50.

Lee S, Charters AC, Chandler JG, Orloff MJ. A technique for orthotopic liver transplantation in the rat. Transplantation. 1973;16(6):664–9.

Oldani G, Maestri M, Gaspari A, Lillo E, Angelastri G, Lenti LM, Rademacher J, Alessiani M, Dionigi P. A novel technique for rat liver transplantation using Quick Linker system: a preliminary result. J Surg Res. 2008;149(2):303–9.

Tan F, Chen Z, Zhao Y, Liang T, Li J, Wei J. Novel technique for suprahepatic vena cava reconstruction in rat orthotopic liver transplantation. Microsurgery. 2005;25(7):556–60.

Tsuchimoto S, Kusumoto K, Nakajima Y, Kakita A, Uchino J, Natori T, Aizawa M. Orthotopic liver transplantation in the rat. A simplified technique using the cuff method for suprahepatic vena cava anastomosis. Transplantation. 1988;45(6):1153–5.

Werker PM, Kon M. Review of facilitated approaches to vascular anastomosis surgery. Ann Thorac Surg. 1997;63(6 Suppl):S122–7.

Zeebregts CJ, Heijmen RH, van den Dungen JJ, van Schilfgaarde R. Non-suture methods of vascular anastomosis. Br J Surg. 2003;90(3):261–71.

Chapter 10
Mouse Skin Transplantation

Weihua Gong

Mouse nonvascularized skin transplant model is used to investigate the immunological mechanisms of graft chronic/acute rejection and mainly characterized by its easy surgical procedure compared with other vascularized transplant models. Importantly, skin graft rejecting process is reliable and reproducible in the aspect of allogeneic responses, offering its wide use in the field of murine transplant researches. Skin grafting has become a basic and established operation technique, applicable to other small lab animals such as guinea pigs and rabbits.

Some features are to be highlighted for the skin transplant model as follows.

1. In surgical technique, thin skin grafts show graft contraction more often than thick skin grafts, although it is unclear why the contraction differs by the thickness of skin grafts. Therefore, the size of donor skin should be slightly bigger than that of graft bed in the recipient.
2. The skin graft can be exchanged between donor and recipient for transplant studies without any sacrifice. One recipient can receive more than two skin grafts, particularly useful for third-party investigation.
3. Immunogenicity: ear skin > back/trunk skin > tail skin. The immunogenicity of skin allograft positively correlates with the density of epidermal Langerhans cells (LCs). Ear skin has the highest density of LCs, followed by trunk skin, and then tail skin, that is say, engraftment with ear skin will be more likely to elicit allogeneic responses compared with the utilization of tail skin (Matsuo et al. 2007).
4. The survival of skin grafts is dependent on the rapid reestablishment of an adequate blood circulation. And the skin graft revascularization is influenced by endothelial and endothelial progenitor cells from the recipient.

W. Gong, MD, PhD
Department of Surgery, The Second Affiliated Hospital of Zhejiang University School of Medicine, Hangzhou, People's Republic of China
e-mail: weihua.gong@gmail.com

W. Gong (ed.), *Rodent Transplant Medicine*,
DOI 10.1007/978-94-017-9472-5_10

10.1 Special Surgical Tool

1. A small hemostat
2. Vaseline
3. Corneal scissors

10.1.1 Preservation Solution

Normal saline and PBS are normally used as preservation solutions. The Roswell Park Memorial Institute (RPMI) as a reliable storage was reported to store human skin graft in nude mouse model, in which antibiotics such as penicillin and streptomycin were suggested to use to control bacterial infection and improve viability. Additional use of fetal calf serum in the preservation solution did not improve human skin viability.

10.2 Surgical Procedure

10.2.1 Donor Surgery

The mouse is anesthetized by intraperitoneal injection of the mixture of Ketamine (100 mg/kg) and (xylazine, 10 mg/kg).

There are three sources of skin graft from the donor mouse including ear skin, trunk/back skin, and tail skin. The difference between these three kinds of skin is as described above. In addition, to ensure the quality of surgical technique, it is important and necessary to have syngeneic transplants as controls.

10.2.1.1 Trunk Skin Surgery

Hairs on the trunk of the donor mouse are shaved and a full-thickness skin is swabbed with 70 % alcohol. The skin is spread on the sterile Petri dish containing normal saline and the superfluous fibrous and fat tissues are cleaned by using corneal scissors, or a number 10 Bard Parker blade, or large gauze sponges. This cleaned skin is washed with normal saline. Only dermal layer is well retained. The whole sheet is be utilized for a couple of recipient due to the large size and each patch of donor skin is trimmed with a scalpel or scissors at the size of 1.0–1.5 × 1.0–1.5 cm for grafting. The donor skin patches are stored in sterile PBS or normal saline until used for transplantation (<30 min). It is important to mark the original direct of the graft, which will help to differentiate survived graft from recipient's skin during posttransplant observation.

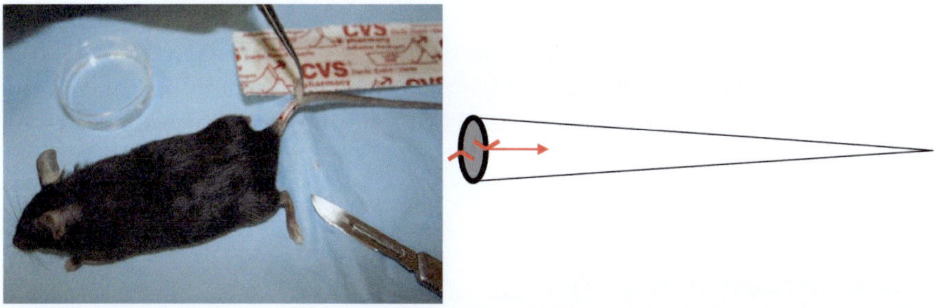

Fig. 10.1 After a skin circumcision and a longitudinal skin incision (1 cm), tail skin is peeled from the donor's tail with a small hemostat or forceps

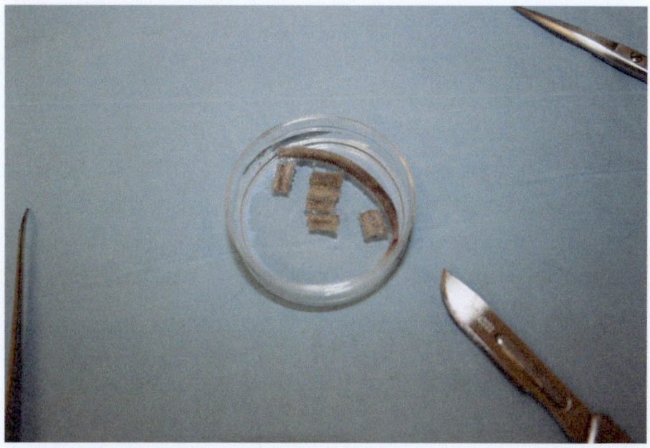

Fig. 10.2 Donor tail skin patches for transplantation

10.2.1.2 Tail Skin Procurement

After the mouse is anesthetized, no hair is to be clipped on tail but the mouse tail should be prepared with 70 % alcohol. A skin circumcision is made at the base of the mouse tail and then a longitudinal skin incision (1 cm) starts from the circumcision (Fig. 10.1). Afterward, the skin can be peeled from the donor's tail with a small hemostat or forceps and immediately spread it in a sterile saline/PBS filled Petri dish. The skin temperature drops to 10 °C within 5 min and to 5 °C within 15 min. The tail skin is then cut into five patches (each approximately 0.8–1.25 × 0.8–1.0 cm) (Fig. 10.2) and preserved in the Petri dish on ice.

10.2.1.3 Ear Skin Surgery

The mouse ears are disinfected with 70 % ethanol and excised at their base using surgical scissors, and then submerged in a Petri dish filled with normal saline. The ears are split into ventral and dorsal halves with a small hemostat or forceps. The cartilage of the relatively hairless ventral halves should be gently scrapped off by using the blunt side of surgical blades prior to implantation. Although the dorsal ear skin has little or none of the cartilage, the adherent fibrous tissue should also be scrapped off before engraftment to the trunk of the recipient. Ear skins approximately sized at 15 × 1 cm are shortly stored in iced saline.

10.2.2 Recipient Surgery

Although nothing but attaching graft on host bed may be applied for implantation, it is not recommended to use this method owing to the difficulty in graft retention. In practice, there are two mainly different approaches to perform engraftment: suturing and tissue-adhesive techniques (TAT).

The recipient mouse is anesthetized as the donor. The hairs around the chest and abdomen are clipped. The mouse is secured on a surgical board. The surgical area is prepared with 70 % alcohol. An incision of the square frame on the lateral thoracic area of recipient mouse is made for the graft bed by using corneal scissors. The skin graft is placed in a slightly smaller graft bed on the recipient. Given the larger size of graft bed, the perpendicular pressure and the difficulty of conglutination will respectively emerge by sutures or tissue adhesive.

10.2.2.1 Suture Technique

The donor skin is grafted onto the center of the dorsal region or the flank of the anesthetized recipient mouse (Fig. 10.3). Two diagonal stitches are firstly done to fix the graft and then other two angles of donor skin are sutured with 5-0 absorbable silk. One or two interrupted sutures are performed for each side of skin graft. Alternatively, a running suture with 6-0 silk (baseball stitch) may be done after skin graft is in place. To keep suitable tension on skin graft, space between stitches on the recipient is slightly bigger than on the graft. The disadvantage of the continuous suturing is that the graft may slip even lose when the silk is broken.

10.2.2.2 Tissue-Adhesive Technique (TAT)

After positioning on its bed, the skin graft is "sealed in place" by tissue adhesive (n-butyl cyanoacrylate, 3 M™ Vetbond) (Fig. 10.4). It is noticeable that the large space between graft and bed can result in a filling of tissue adhesive, affecting revascularization and survival of skin graft. Therefore, the graft is gently pressed and firmly against the bed before the application of the tissue adhesive. In addition, Vaseline implementation will not perform until tissue adhesive becomes dry.

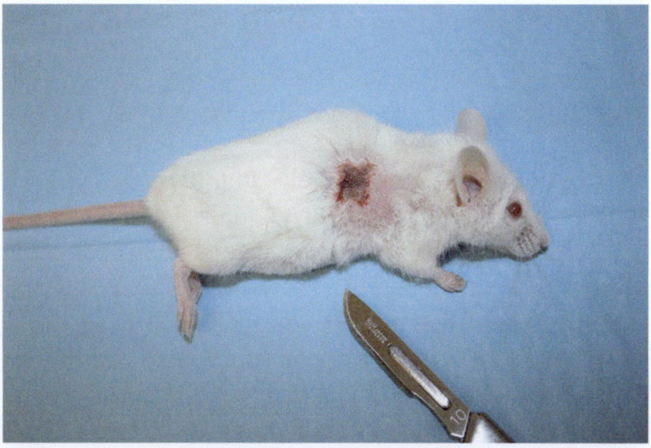

Fig. 10.3 Donor skin patch is grafted onto the center of the flank of recipient mouse

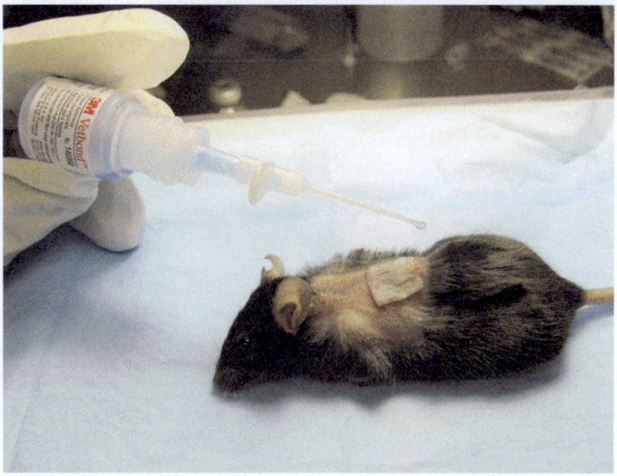

Fig. 10.4 The skin graft is "sealed in place" by tissue adhesive (Courtesy of Dr. Caroline Ettinger from the University of Regensburg, Germany)

10.3 Dressing After Surgery

To avoid dislodgment of skin graft, the recipient is bandaged with a nonadherent Vaseline-impregnated gauze and further protected by using a bandage and paper tape. Alternatively, given that teeth and claws of the recipient are trimmed, no any dressing is required after surgery.

10.4 Disbandagement

The bandage and dressings are removed using bandaging scissors 5–7 days after transplant surgery. In the first week, the recipient or its cagemates can scratch or bite off the skin graft owing to the skin sensitivity or itch. However, with regard to accelerated acute or acute rejection model, the bandage should be removed earlier for close observation.

10.5 Observation of Graft Survival

The grafts are visually monitored and scored daily on basis of the percentage of necrotic skin area in planimetric scales. The long-term surviving grafts can be inspected daily for first 2 weeks and then observed weekly. The healthy skin graft shows soft and pliable to touch and the hairs on the graft regrow. As the skin graft is rejecting, the dermal graft tissue becomes necrotic and scab. The host epidermal cells replace it thereafter. Therefore, graft is considered fully rejected when it becomes completely necrotic or scabbed. With respect to chronic rejection, the graft is evaluated according to the proportion of healthy hair follicles and the rejection endpoint is determined by collagen pad formation, loss of more than 90 % of their hair, and severe contraction of skin graft. Conversely, the hair on the graft remains growing for long-term survival graft in different direction when the graft piece is intentionally positioned in opposite direction with the recipient's hair.

Indeed, the blind determination of a complete rejection (score 5) is the critical point of mouse skin transplant as direct vision is in a large measure arbitrary. Thereby, as electronic scientific technique is progressing, analysis based on digital photography (camera) has been applied to monitor allograft rejection. In particular, the early graft rejection can be clearly assessed by magnifying the graft through a microsurgical microscope. Serial pictures for each mouse will be assigned to a separate folder in chronological order. The rejection extent is defined by the percentage of necrotic part in total graft area. The following score system may be used to record the rejecting process: score 0 means intact graft; 1 = onset of graft rejection, <25 % rejection or shrinkage; 2, >25 % rejection or shrinkage; 3, >50 % rejected or over 50 % shrinkage of its original size; 4, >75 %; and 5, complete rejection (>95 %). To avoid different scoring by person, the assessment should be independently scored by additional two observers. Noticeably, although rejection is complete, the graft area is extremely rapidly replaced by host epithelium.

10.6 Histological Evaluation of Skin Graft

The harvested skin allo- or isografts are analyzed by conventional histological H&E staining and immunohistochemistry including the evaluation of the structures of the epidermis, the dermis, hair follicles, nerves, blood vessels, and sebaceous glands. All slides are evaluated by a dermatopathologist in a blinded fashion.

10.7 Analgesic

Buprenorphine at a dosage of 0.05 mg/kg is intramuscularly given very 12 h for 48 h to control postoperative pain.

10.8 Admission of Graft for Analysis

The grafts slipped or shriveled in the first week should not be included for study analysis.

10.9 Discussion

Skin transplantation is a basic and well-established surgical technique as it is an easy and quick method for studying transplant biology. Skin transplantation can be used to investigate the efficacy of various immunosuppressive drugs and assess tolerance induction protocols. Herein, a couple of issues of surgical techniques and relevant immunobiology are reviewed to better understand skin transplantation.

By far, there are following surgical techniques adopted for skin transplantation: suturing (continuous or interrupted), dressing, glue, and nothing. The technique of continuous suturing was described for skin transplantation by Dr. Hardin in 1954, by which full-thickness body skin was transplanted into recipient's dorsal body area (Sugarbaker and Chang 1979). Currently, interrupted suturing is normally used as we shown here. As past studies mentioned before (Sugarbaker and Chang 1979), an attempt was made by us to test skin transplantation without any suture or dressing. The findings showed that the technique using uncovered skin grafts was not reliable. The adherence of skin graft and bed was not satisfying and some skin grafts dislodged from their beds. One obvious technical benefit of the dressing was that it totally prevented grafts from becoming dislodged from their beds. It was reported that no slippage of the more than 1,500 murine skin grafts transplanted by this procedure has ever occurred (Baldwin et al. 1973). Since fibrin glue could enhance the adherence of skin graft to bed, it was suggested to be superior to suture technique (Brown et al. 1992). Nevertheless, use of glue causes increase of experimental cost. And it is obscure whether fibrin glue will interfere with the process of immunoresponses in graft bed. By far, suturing technique remains common method for experimental studies. Irrespective of any techniques, wrap bandage is routinely used to cover the transplanted skin graft to avoid loss by other mice biting in the same cage. Caution should be taken to suitably wrap the bandage to keep breathing. Tight wrapping may stop breathing and loose wrapping may cause loss of bandage. Continuous inspection of skin graft should be documented and scored by using digital photography and the recipient with skin graft fall is excluded from experimental studies (Schwoebel et al. 2005).

After skin transplant surgery, skin graft adapts to host's bed and replaces their biological function. This entire process was interestingly unveiled by using green fluorescent protein (GFP) transgenic mice. The findings revealed that the tissues from ectodermal origin of donor including hair follicles, the epidermis, and sebaceous glands survived in the recipient's bed, and that some parts of the dermis, nerves, and blood vessels from recipient also survived in the transplant bed

(Matsuo et al. 2007). Although skin transplantation is not vascularized, chilling of graft is necessary for better graft survival (Garrod and Cahalan 2008).

Notably, skin allografts from ears, tails, and trunk might evoke different degree of immune responses. It was easier to elicit strong allogeneic immunoresponses by transplanting full-thickness back skin compared to tail skin as more epidermal Langerhans cells (antigen-presenting cells) reside in the former (Matsuo et al. 2007). Indeed, graft from ear skin has a higher density of LC compared to trunk or tail skin (Larsen et al. 1990). The size of Langerhans cells (LC) could dramatically increase in donor's epidermal sheets and their expression of MHC II molecules were also augmented, although their amount might remarkably decreased (Larsen et al. 1990). In vivo experimental data exhibited that epidermal LC migrates into lymphatics via the dermis and then to the draining lymph nodes, where T cells are homing. This process provides a creation of a powerful stimulus for allograft rejection (Larsen et al. 1990). Herein, tissue characteristics of donor skin per se influence allorecognitions.

The tail skin graft of Balb/c recipient receiving donor skin from gender-matched and fully MHC-mismatched C57BL/6 (H2b) mice is normally rejected at day 7–10 (Schwoebel et al. 2005). Skin allografts from trunk (C57BL/6→Balb/c) can survive for 8–14 days without any treatment (Li et al. 2008). However, 70 years ago, Dr. Bilingham RE et al. found that MHC-mismatched mouse skin allografts could permanently survive if transplant recipients previously underwent a exposure of donor antigen during the neonatal period (Billingham et al. 1953).

References

Baldwin 3rd WM, Cohen N, Hrapchak BB. Prolonged survival of murine skin grafted across a weak histocompatibility barrier as a function of skin-grafting technique. Transplantation. 1973;15(4):419–22.

Billingham RE, Brent L, Medawar PB. Actively acquired tolerance of foreign cells. Nature. 1953;172(4379):603–6.

Brown DM, Barton BR, Young VL, Pruitt BA. Decreased wound contraction with fibrin glue–treated skin grafts. Arch Surg. 1992;127(4):404–6.

Garrod KR, Cahalan MD. Murine skin transplantation. J Vis Exp. 2008;11:e634.

Larsen CP, Steinman RM, Witmer-Pack M, Hankins DF, Morris PJ, Austyn JM. Migration and maturation of Langerhans cells in skin transplants and explants. J Exp Med. 1990;172(5): 1483–93.

Li QY, Chi YY, Liu SQ. Cell cycle arrest effects of large-dose FTY720 on lymphocytes in mouse skin transplantation models. Immunopharmacol Immunotoxicol. 2008;30(2):365–81.

Matsuo S, Kurisaki A, Sugino H, Hashimoto I, Nakanishi H. Analysis of skin graft survival using green fluorescent protein transgenic mice. J Med Invest. 2007;54(3–4):267–75.

Schwoebel F, Barsig J, Wendel A, Hamacher J. Quantitative assessment of mouse skin transplant rejection using digital photography. Lab Anim. 2005;39(2):209–14.

Sugarbaker PH, Chang AE. Uncovered skin grafts in mice. J Immunol Methods. 1979;31(1–2): 167–75.

Chapter 11
Mouse Heterotopic Abdominal Heart Transplant Model

Weihua Gong

11.1 Introduction

Currently, more than 1,300 strains of genetically modified mice have been developed to better understand signaling pathways and immunobiology. In the field of transplantation, utilization of these strains is not sufficient to achieve the goal of advance of transplant biology, but a prerequisite for a success. Implementation of various mouse transplant models is additionally required during the past decades, which offers a valuable platform to unveil the complexity of the immune system and clinically relevant transplant issues on basis of a high degree of similarity between human leukocyte antigen (HLA) and mouse H-2 MHC complex (Erickson 1996).

As Drs. Corry and Russell firstly reported mice abdominal heterotopic heart transplantation model in 1973 (Corry et al. 1973), it has been widely adopted by a considerable number of international research centers owing to its vascularized transplantation, relatively simplicity of this model (compared to mouse liver and kidney transplantations), and advantages in viably explaining transplant biology. In fact, heterotopic cardiac transplant models comprise of abdominal and cervical different techniques based on distinct grafting location. Cervical model was firstly established by Dr. Zhonghua Chen in order to further test donor organ specific tolerance on the same recipient by using another heart graft from third party (Chen 1991).

Abdominal heterotopic heart transplant technique is conventional and widely used for transplant studies as it is characterized by the following: (1) large surgical area and (2) big recipient's aortic/inferior vena cava vessels for two end-to-side anastomoses with donor ascending aorta and pulmonary artery.

W. Gong, MD, PhD
Department of Surgery, The Second Affiliated Hospital of Zhejiang University School of Medicine, Hangzhou, People's Republic of China
e-mail: weihua.gong@gmail.com

W. Gong (ed.), *Rodent Transplant Medicine*,
DOI 10.1007/978-94-017-9472-5_11

11.2 Mice

Inbred wild-type (WT) C57BL/6 (H-2^b) and BALB/c (H-2^d) male mice at 8–12 weeks of age are used for experimental studies and housed under pathogen-free facilities. Research protocols are approved by the Institutional Animal Care and Use Committee (IACUC). Allogeneic heterotopic heart transplants are performed by using BALB/c (H-2^d) donors and C57BL/6 (H-2^b) recipients; syngeneic cardiac transplantations are accomplished by using C57BL/6 (H-2^b) donors and C57BL/6 (H-2^b) recipients.

11.3 Surgical Instruments

All the microsurgical procedures for mouse heart transplantation are accomplished under a desk-top operating microscope at a magnification of 4× to 25× (Carl Zeiss Operating Microscope, f170, Opmi pico, Germany). Microsurgical instruments include straight supergrip forceps, microserrefine clamps, a curved needle holder (World Precision Instruments, USA), and microdissecting scissors (Fine Science Tools, Germany). 10-0 swaged microsurgical suture (Fine Science Tools) is adopted for vascular anastomosis. Retractors that made by clips are used to expose the surgical area. All these instruments are autoclaved to disinfect. A low-temperature Cautery (World Precision Instruments) is used to coagulate recipients' small vessels such as tributaries of the external jugular vein. The Cautery is disinfected by using 70 % alcoholic.

11.4 Anesthetic

In fact, there are two different ways for anesthetizing mice including inhalation and injection.

1. Isoflurane is normally inhaled at a concentration of 1–3 % in oxygen through a precise vaporizer after initial induction at 4–5 % in an induction chamber. Advantage of inhalation of isoflurane is rapid induction and recovery. The main disadvantage is that the equipment is costly.
2. Ketamine (100 mg/kg) in combination with alpha2-agonists (xylazine, 10 mg/kg) are injectable anesthetics for surgeries. The advantage of ketamine is its wide margin of safety. The combined use of ketamine and xylazine (mixture, intraperitoneally) can maintain a surgical plane of anesthesia for about 30 min. With regard to heterotopic heart transplant, moisture gauge is provided for covering visceral organs and cold saline is used. These maneuvers decrease mice body temperature, leading to an enhanced effect of anesthetic mixture. If surgery is in need of more time, the mixture can be sprayed onto visceral organs. The major disadvantage is that ketamine is a Class-III controlled drug and a license is required for use.

Before incision surgery is made, rear foot reflexes should be tested to reveal the status of anesthesia.

11.5 Storage Solution for Donor Organ

Normal saline (sodium chloride 0.9 %), lactated Ringer's solution (a standard isotonic solution contains 6.5 g NaCl, 0.42 g KCl, 0.25 g $CaCl_2$, and 1 mol of sodium bicarbonate in one liter distilled water), histidine-tryptophan-ketoglutarate (HTK) solution (a high-flow and low-potassium preservation solution), and University of Wisconsin (UW) solution are commonly used as storage/perfusion solution in transplant surgery.

It is widely accepted that HTK and UW solutions are supreme to normal saline and Ringer's, offering transplant organ better protective effect. HTK was originally introduced as a cardioplegic solution by Dr. Bretschneider (from Germany) in the 1980s and used as organ storage solution since 2002. In comparison to UW solution, HTK has lower viscosity and is less costly. Indeed, based on our own experience, the protective effect of HTK and UW solution is equivalent for 8–12 h organ preservation. The vast difference of efficacy of these two solutions arises as the preservation time exceeds 24 h. UW is evidently advantageous over HTK for long-term preservation.

11.6 Mouse Abdominal Heart Transplant Model

11.6.1 Donor Heart Preparation

Under deep anesthesia, the surgical area of donor is prepared by shaving hair, disinfecting with Betadine solution and 70 % alcohol. After a long, midline abdominal incision was made, 0.3-ml saline solution with heparin (50–100 U) was injected into the inferior vena cava (IVC). Two minutes later, transaction of abdominal aorta and IVC is done to release circulating blood. Afterward, a U-shaped bilateral thoracotomy is performed to expose heart. Small pieces of ice instead of cardioplegic solution (potassium-enriched solution) (Chen 1991) are immediately applied to cover beating heart in order to arrest its beating and minimize formation of the toxic end products of anaerobic metabolism within donor organ. The superhepatic IVC close to right atrium is ligated with 5-0 or 6-0 silk suture (Fig. 11.1). Afterward, the ascending aorta (Fig. 11.2B) and pulmonary artery (Fig. 11.2C) are dissected and transected at the level of bifurcation of pulmonary artery. The superior vena cava (SVC) is ligated (Fig. 11.2A). The remaining vessels are tied together with 5-0 or 6-0 silk suture. The heart is lavaged with precooling heparinized saline solution (50 U/ml) (4 °C) through the ascending aorta (Fig. 11.3) and perfusate is drained out through pulmonary artery. The harvested graft is stored in a Petri dish containing with cold preservation solution.

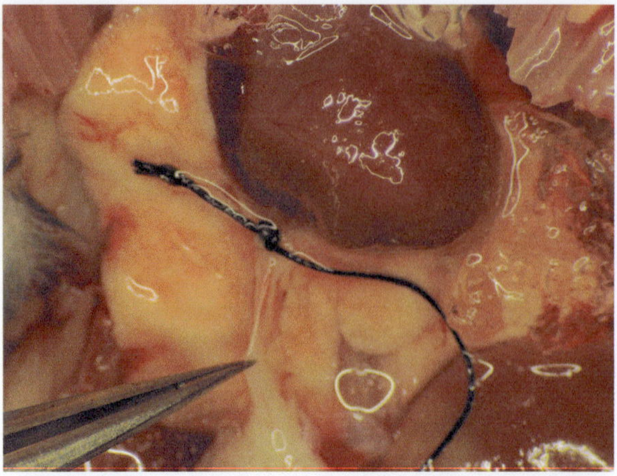

Fig. 11.1 The super-hepatic IVC close to right atrium is ligated with 6-0 silk suture

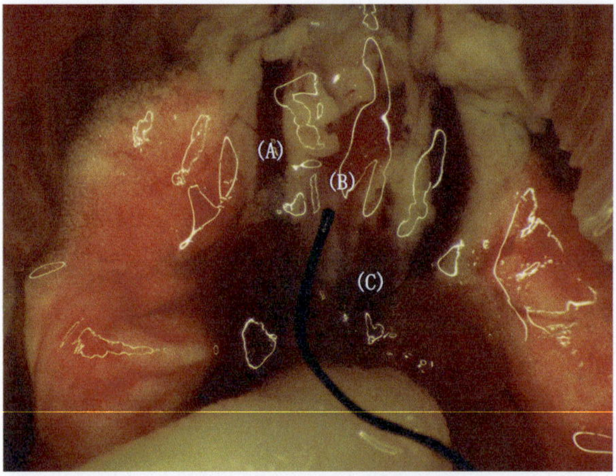

Fig. 11.2 The ascending aorta (*B*) and pulmonary artery (*C*) are dissected and transected at the level of bifurcation of pulmonary artery. The superior vena cava (SVC) (*A*) is ligated

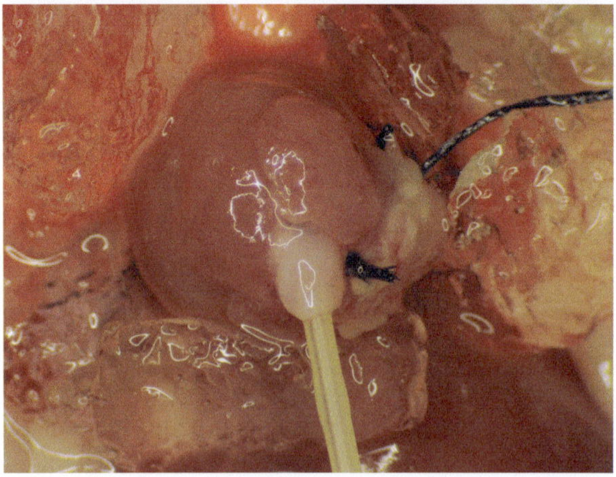

Fig. 11.3 The heart is lavaged with precooling heparinized saline solution (50 U/ml) (4 °C) through the ascending aorta

11.7 Recipient Surgery

After the recipient mouse is anesthetized, abdominal surgical area is prepared by shaving hair and disinfecting with Betadine and 70 % alcohol solution. The mouse is properly positioned supine on the surgical board with the caudal part of its body toward the microsurgeon. A midline laparotomy was performed and the abdominal cavity was clearly exposed by using self-made retractors. The intestines are taken out from the peritoneal cavity, placed on the top of anterior chest and surrounded by moisture gauze. IVC and abdominal aorta between left renal vessels and bifurcation of aorta are dissected. Care should be taken to avoid damaging any nerves as the lumbar veins are cauterized. Two ties with 6-0 silk sutures or two microserrefines (straight, 15-mm length, 100 g pressure, and 6 mm×1 mm jaw dimensions) are placed in the distal and proximal parts of IVC and abdominal aorta to temporarily interrupt the blood flow for subsequent anastomosis.

11.8 Arteriotomy and Venotomy

The direct puncture or longitudinal excision with sharp blade is not recommended to use for arteriotomy owing to their high incidence of anastomotic obstruction posttransplant. Instead, the following two techniques are suggested to successfully apply: (1). The head of 10-0 suture needle (5.0 mm, 3/8c) is longitudinally passed in and out of the anterior median wall (AMW) of the abdominal aorta and is kept in place. While the head part of the needle is slightly lifted up by holding the tail part with the forceps, in parallel, the lifted AMW is transected to make an elliptical aortotomy by excising beneath the head part of the needle with the sharp tip of Vannas spring scissors. (2). An elliptical arteriotomy is made by cutting with an iris scissors under the guidance of a10-0 suture instead of the above-described 10-0 needle in the AMW of the abdominal aorta and a longitudinal venotomy is directly pierced with an iris scissors (Fig. 11.4). The size of the opening in the recipient's IVC could be equivalent to or a bit bigger than the diameter of the donor's IVC, while the arteriotomized size could be the same or a bit smaller than that of the donor's ascending aorta. Noticeably, given that recipient's opening is formed too larger, oblique transaction can be made for donor's vessel to fit the opening. The thrombus in the blocked vessels should be thoroughly flushed out by pre-cooling heparinized saline solution (10 U/ml) in order to prevent thrombosis after surgery. During the process of these preparations, it is necessary to protect vascular intima from any impairment and meticulous manipulations are required.

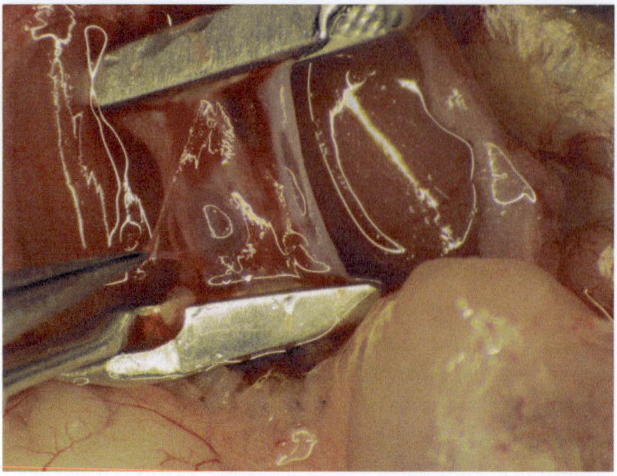

Fig. 11.4 An elliptical arteriotomy is made by cutting with an iris scissors under the guidance of a 10-0 suture in the abdominal aorta and a longitudinal venotomy is directly pierced with an iris scissors

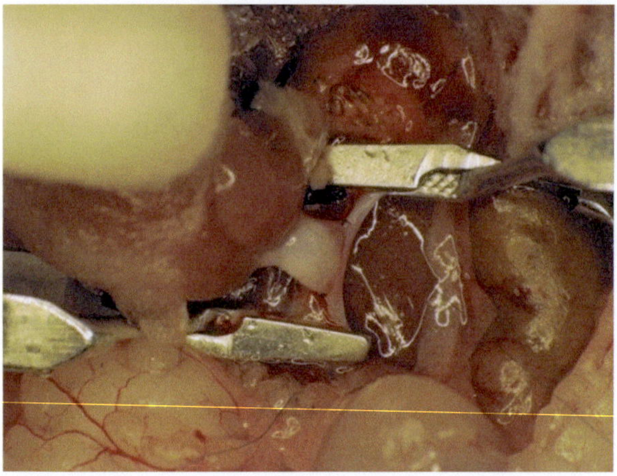

Fig. 11.5 Two stay sutures with 10-0 nylon are performed at the proximal and distal apexes of the opening of abdominal aorta to anchor the graft

11.9 Vascular Reconstruction

The heart graft is gently positioned into right side of the recipient's abdominal cavity, indicating warm ischemia starts until blood flows through graft via accomplished vascular reconstruction. The graft is wrapped by iced saline soaked cottage to prevent graft from slow-warming injury by the microscope and the recipient's body temperature. Two stay sutures with 10-0 nylon are performed at the proximal and distal apexes of the opening of abdominal aorta to anchor the graft (Fig. 11.5), which will facilitate the anastomosis progress. Afterward, the end-to-side anastomoses are carried out by using 10-0 nylon suture. Running sutures starts from the upper stay stitch and are done in a clockwise direction. Three to four stitches are applied for each side based upon the size of opening (Fig. 11.6). As left side of

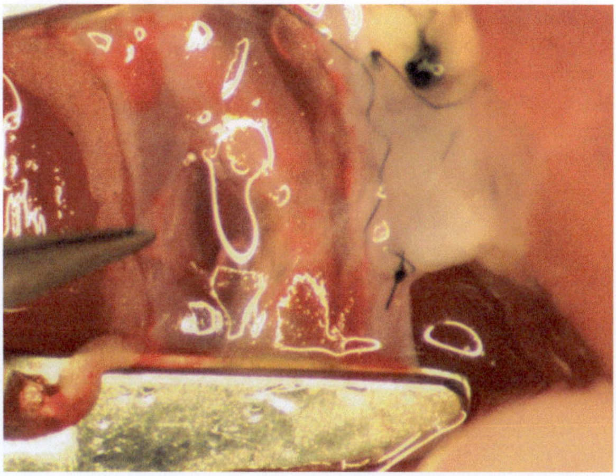

Fig. 11.6 Running sutures starts from the upper stay stitch in a clockwise direction. Three to four stitches are applied for each side based upon the size of opening

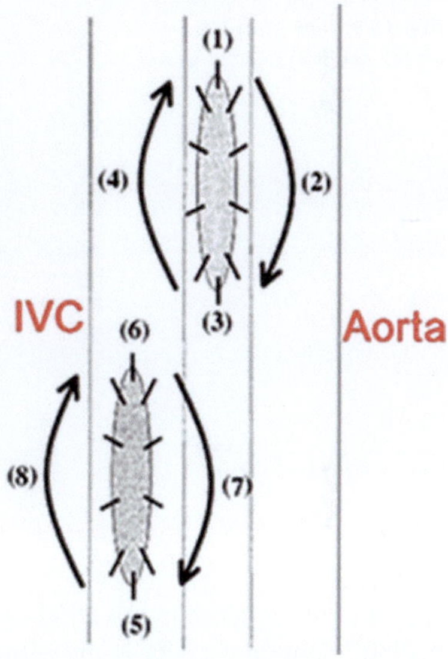

Fig. 11.7 Running sutures are given for venous and arterial reconstruction in a clockwise direction (Modified on basis of Niimi (2001)). This direction of running suturing is instructive in better exposing operational area, decreasing movement frequency of graft, and shortening operation time.

artery anastomosis is accomplished, the graft will be gently moved to the left abdominal cavity for suturing right side of the opening in a similar fashion (Fig. 11.7). The last stitch was done by a knot with the first stay stitch. It is important to point out that a tiny space should be kept for last stitch. Otherwise, the tight anastomosis will result in poor perfusion.

With respect to venous reconstruction, the anterior wall of the IVC is gently lifted with forceps to avoid penetrating the venous wall during stitching. Importantly, it is also of great help in clearly exposing the anastomotic site and distending donor pulmonary artery by flushing them with cold preservation saline. Thereafter, distal and proximal stay sutures are made but the distal one remains untied for better

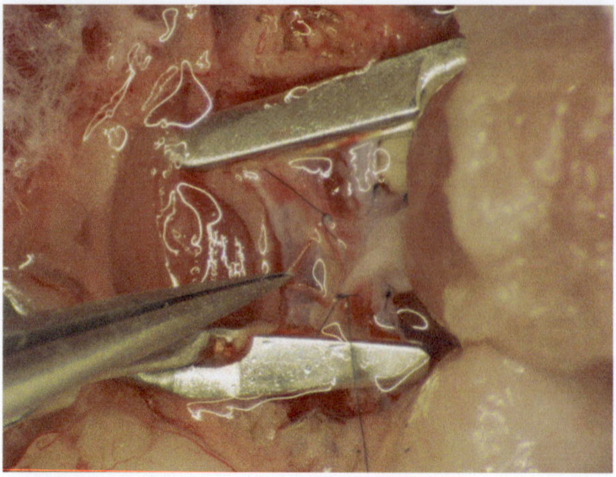

Fig. 11.8 The left side of venous anastomosis is accomplished by in-out stitches. The right side of anastomosis is stitched by out-in pattern

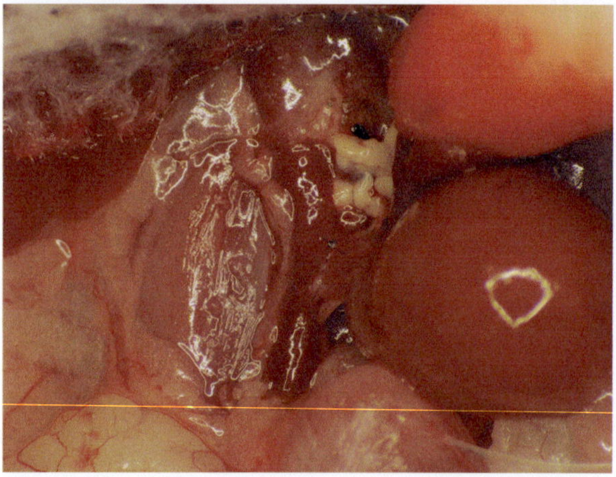

Fig. 11.9 The graft is reperfused and flushed in *red*

exposure during continuous stitches for venous anastomosis. As the graft remains in the left abdominal cavity in the entire course of venous anastomosis without turning the heart over, the first stitch of running sutures is adopted as an out-in pattern and others in the left side are done by in-out stitches. The suture needle is then passed through the distal apex of the opening. Subsequently, the right side of anastomosis is stitched by out-in pattern (Fig. 11.8). The last stitch was done by a knot with the proximal stay stitch. During the surgery, cool saline can be repetitively sprayed onto the graft to keep coldness. After inflow and outflow tracts of graft are successfully reconstructed, the distal microvascular clamp is firstly released owing to the lower pressure from the IVC. In parallel, the proximal one is also removed immediately and then the graft is reperfused, flushed (Fig. 11.9). The atriums of heart graft firstly recover and begin to beat. Sometimes, a short period of ventricle fibrillation of the graft take place, followed by sinus rhythm. It is important to record the following

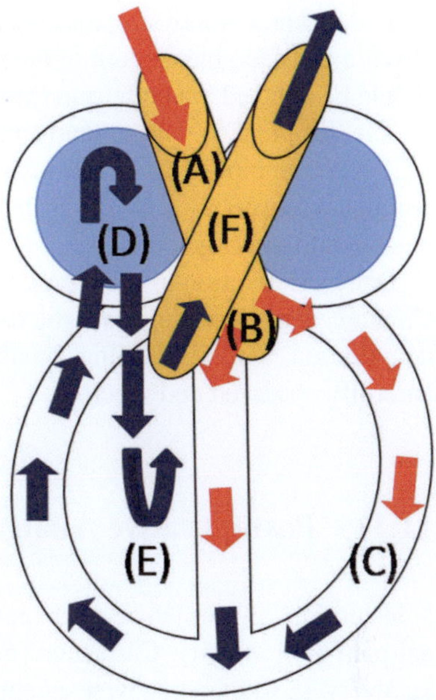

Fig. 11.10 This diagram is designed to describe the blood flow after vascular reconstruction in this transplant model. (*A*) The ascending aorta, (*B*) origin of coronary artery, (*C*) myocardium, (*D*) right atrium, (*E*) right ventricle, and (*F*) pulmonary artery

information: cold ischemia time (time period between the start of donor organ perfusion and the end of organ preservation in cold solution), warm ischemia time (starting from the end of organ preservation in cold solution to blood reperfusion), and rebeating time (starting from the blood reperfusion to fully heart beating).

The intestines are moved back to the abdominal cavity and should be put in order to avoid intestinal obstruction. Two layers of the abdomen are closed separately with continuous 4-0 or 5-0 silk/absorbable sutures.

Like other visceral organs, this transplant graft in the abdomen is not de facto pumping blood as the pattern of blood circulation is different from the normal. In detail, the blood from the recipient's abdominal aorta initially flows into the ascending aorta of the transplanted graft and then enters the coronary arteries. The blood supplies the entire heart graft and is drained into the right atrium through the coronary sinus. Furthermore, the blood passes through right ventricle and leaves graft for the recipient's IVC via the donor's pulmonary artery (Fig. 11.10).

11.10 Postoperative Care

As heterotopic heart transplantation in the abdomen is a relatively large surgery for mouse, intraoperative fluid and blood loss is evitable. Furthermore, heating from the microscope lamplight could accelerate the process of fluid loss. Therefore, warm normal saline (0.5–2 ml) should be administered subcutaneously and the fluid volume is dependent upon mouse size, total operation time, blood loss, and intensity of lamplight. Given a considerable amount of blood loss during surgery, warm physiological saline should be injected via body vein (penis vein, the IVC, and the tail vein).

The recipient is subsequently placed in a microisolator cage. Half part of the cage with the mouse should be positioned on the circulating water blanket until the recipient completely recovers from the anesthesia. Juice gel and wet ad libitum are placed onto the cage button.

Daily observation of the recipient's health status and graft function is suggested. Clinical signs including weight loss, respiration rates, consciousness, cyanosis, dehydration, wound bleeding, breathing, moving, paralysis, and activity should be carefully observed and record.

11.11 Postoperative Analgesia

Nonsteroidal anti-inflammatory analgesics should be used for alleviating any residual pain after surgery. Carprofen (6 mg/kg, daily) or buprenorphine (0.05–0.1 mg/kg, every 12 h) may be given subcutaneously for the 48 h posttransplant. Indeed, in terms of injection schedule, the former is more practical and preferable.

11.12 Graft Assessment

Graft function is normally monitored by daily manual palpation of the beating graft. Rejection was defined by the last day of the complete cessation of heart contraction and verified by direct visualization or autopsy. Although the electrocardiogram (ECG) may detect electrical impulses of graft, monitoring heart beating via ECG is still controversial. Graft survival is subjected to the log-rank test or the Gehan-Breslow-Wilcoxon test for statistical comparison.

In fact, the intensity of heart palpation can be graded at a scale of 0–4. Grade 4 indicates normal strong heating, 3 for homogeneous ventricular contraction, 2 for partial or weak ventricular motion, 1 for minimal ventricular contraction, and while 0 represents absence of any contraction (Amberger et al. 2002; Fischbein et al. 2002). With respect to histological evaluation, formalized graft is embedded in paraffin, sectioned, and stained with Gill's hematoxylin and eosin to relatively quantitatively evaluate cellular infiltration, vasculopathy, and the degree of rejection by light microscopy. The severity of vasculopathy can be scored for those vessels with clear internal elastic lamina on basis of the percentage of intimal thickening and luminal occlusion of vessels. Score 0 represents a normal artery, 1 for 10–20 % vascular luminal occlusion, 2 for 20–50 % luminal occlusion, and while score 3 indicates more than 50 % luminal occlusion (Yang et al. 2008). It is noteworthy that all the histological analysis should be blindly performed to ensure objectivity of the evaluation.

11.13 Admission Standard of Transplanted Recipients

The heart graft that does not beat within 48 h posttransplant should be considered a technical failure and should be excluded from further analysis regardless of any cause.

11.14 Discussion

Abdominal heart transplant model is the most frequently used for mouse transplant studies, in which donor pulmonary veins and inferior vena cava are ligated. Donor pulmonary trunk and ascending aorta are anastomosed into recipient's IVC and abdominal aorta respectively. This model's limitation is absence of pumping function of heart graft although the direction of coronary blood flow is normal and myocardial contractility is preserved. However, the pressure of coronary artery may be different from physiological condition since blood is retrograded into donor's ascending aorta with high pressure.

Nevertheless, abdominal heart transplant model is commonly used to investigate transplant biology, particularly for murine transplantation. Different combinations of mice for various organ transplantations have been widely described (see Chap. 1), which offers us a well platform for experimental studies such as chronic rejection (Bedi et al. 2010) or acute rejection of heart graft (Gong et al. 2012). To achieve acute rejection, combination of Balb/c and C57BL/6 can be utilized to explore impact of immunologic factors on allograft injury. In this model, allograft rejection normally appears between day 7 and day 10 without any treatment after transplant surgery, as evidenced by notable accumulation of intramuscular and pericardial mononuclear leukocytes and cessation of heart beating (Gong et al. 2012). Development of chronic rejection model can unveil the underlying mechanism of chronic allograft injury and long-term graft dysfunction (Bedi et al. 2010). Irrespective of different progress of rejection, our experimental observations displayed that pericardial inflammation severely appeared particularly in the atrium part, whereas the cardiac apex was weak or even absent.

References

Amberger A, Schneeberger S, Hernegger G, Brandacher G, Obrist P, Lackner P, Margreiter R, Mark W. Gene expression profiling of prolonged cold ischemia and reperfusion in murine heart transplants. Transplantation. 2002;74(10):1441–9.

Bedi DS, Riella LV, Tullius SG, Chandraker A. Animal models of chronic allograft injury: contributions and limitations to understanding the mechanism of long-term graft dysfunction. Transplantation. 2010;90(9):935–44.

Chen ZH. A technique of cervical heterotopic heart transplantation in mice. Transplantation. 1991;52(6):1099–101.

Corry RJ, Winn HJ, Russell PS. Primarily vascularized allografts of hearts in mice. The role of H-2D, H-2K, and non-H-2 antigens in rejection. Transplantation. 1973;16(4):343–50.

Erickson RP. Mouse models of human genetic disease: which mouse is more like a man? Bioessays. 1996;18(12):993–8.

Fischbein MP, Yun J, Laks H, Irie Y, Fishbein MC, Bonavida B, Ardehali A. Role of CD8+ lymphocytes in chronic rejection of transplanted hearts. J Thorac Cardiovasc Surg. 2002;123(4):803–9.

Gong W, Thornley T, Whitcher GH, Ge F, Yuan S, Liu DJ, Balasubramanian S. Introduction of modified cervical cardiac transplant model in mice. Exp Clin Transplant. 2012;10(2):158–62.

Niimi M. The technique for heterotopic cardiac transplantation in mice: experience of 3000 operations by one surgeon. J Heart Lung Transplant. 2001;20(10):1123–8.

Yang J, Popoola J, Khandwala S, Vadivel N, Vanguri V, Yuan X, Dada S, Guleria I, Tian C, Ansari MJ, Shin T, Yagita H, Azuma M, Sayegh MH, Chandraker A. Critical role of donor tissue expression of programmed death ligand-1 in regulating cardiac allograft rejection and vasculopathy. Circulation. 2008;117(5):660–9.

Chapter 12
Mouse Heterotopic Cervical Heart Transplant Model

Weihua Gong and Guangxiang Liu

12.1 Introduction

Although abdominal heart transplantation was earlier developed, transplantation in cervical part was subsequently created for its apparent advantages, including convenient observation of heart-beating and less microsurgical challenges by using cuff techniques (Chen 1991). Indeed, cervical heart transplant models can be accomplished by two different techniques: cuff and suture (Gong et al. 2012). To shorten surgical time and reduce the surgical difficulties, the external jugular vein and the common carotid artery of recipients are normally transected and cuffed for subsequent reconstruction of blood flow, however, which disrupts the blood supply to the central nervous system. The hemodynamical alteration may create high blood pressure in the transplanted graft. On the other hand, the cuff technique has a potential risk of thrombosis, affecting the achievement of long-term graft survival (Heron 1971).

Herein, a novel cervical cardiac transplant model is introduced to establish by preserving the continuity and function of the external jugular vein and carotid artery.

W. Gong, MD, PhD (✉)
Department of Surgery, The Second Affiliated Hospital of Zhejiang University School of Medicine, Hangzhou, People's Republic of China
e-mail: weihua.gong@gmail.com

G. Liu, MD, PhD
Department of Urology and Kidney Transplant Center, Nanjing Drum Tower Hospital, The Affiliated Hospital of Nanjing University Medical School, Jiangsu, People's Republic of China

W. Gong (ed.), *Rodent Transplant Medicine*,
DOI 10.1007/978-94-017-9472-5_12

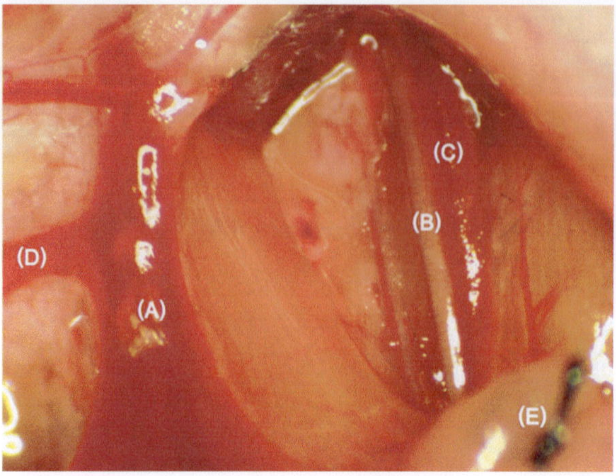

Fig. 12.1 The tributaries (*D*) of the external jugular vein (*A*) are exposed. After the clavicular head (*E*) of the right sternocleidomastoid muscle is excised, the right carotid artery (*C*) and vagus (*B*) nerve are exposed

12.2 Donor Surgery

The donor mouse is anesthetized with ketamine/xylazine mixture and receives a midline incision in the abdomen. 0.3-ml heparinized saline (50–100 U/ml) is injected via the inferior vena cava (IVC). Two minutes later, transaction of abdominal aorta and IVC is done to release circulating blood and then an up-end U-shaped anterior thoracotomy is performed to clearly expose the beating heart. Subsequently, small pieces of ice are immediately applied into the thoracic cavity to cover heart for cardioplegia. The intrathoracic IVC is bluntly dissected and transected in close vicinity to the diaphragm. Afterward, the ascending aorta is freed from the surrounding soft tissue and then incised close to the innominate artery. All other vessels (the pulmonary vessels, the left/right superior vena cava) are ligated together by using a 6-0 silk suture. The en bloc heart graft is then mobilized and preserved in normal saline solution at 4 °C.

12.3 Recipient Operation

The recipient is anesthetized as described and laid supine with the caudal part toward the surgeon and the limbs are held in place with a piece of adhesive tape. Hair is shaved and the surgical field of skin is cleaned with iodine and 70 % ethanol solution. An oblique longitudinal incision is made from the sternum to the right mandibular angle. In this model, the right submaxillary gland, the external jugular vein, and right carotid are kept intact in order to attempt to keep normal physiology and maintain adequate blood flow to the central nervous system. The tributaries (D) of the external jugular vein (A) (Fig. 12.1) are carefully electrocauterized with pen-tip cautery instead of using ligatures, minimizing surgical time. The clavicular head (E) of the right sternocleidomastoid muscle (Fig. 12.1) is excised and mobilized as far as possible for better exposure of the right carotid artery (C). The distal and proximal parts of carotid artery are releasably tied with 6-0 silk sutures to spare

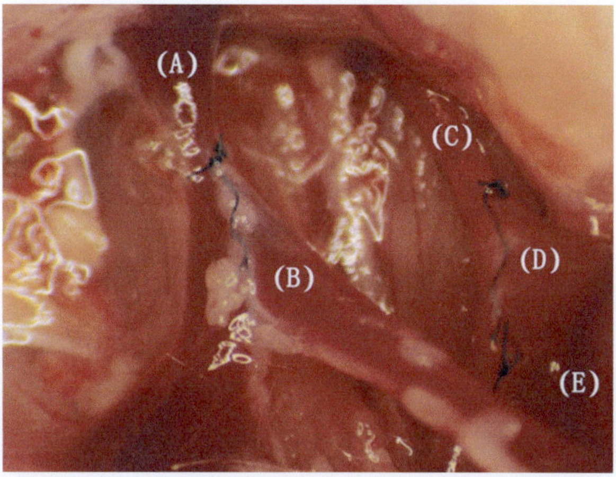

Fig. 12.2 Thedonor's intrathoracic IVC (*B*) and ascending aorta (*D*) are respectively anastomosed end to side to the external jugular vein (*A*) and the right carotid artery (*C*). The graft heart (*E*) is positioned onto the left side of the recipient

enough spaces for subsequent surgical maneuvers, whereas their counterparts of the external jugular vein are clamped to clearly expose the opening and temporarily block blood flow. Both arteriotomy and venotomy are done as aforementioned. Both the carotid artery and the external jugular vein are irrigated with normal saline to clear intraluminal blood and clots.

The donor heart is surrounded with an ice-cold saline pad. Warm ischemia time starts when the graft is positioned in the right flank of surgical area. After two stay stitches are performed to anchor the graft at the distal and proximal apexes of the opening of the carotid artery, end-to-side anastomosis of the donor aorta with the carotid artery of the recipient is continuously sutured with 10-0 nylon in a clockwise direction. Afterward, the donor heart is flipped from the right side of the recipient to the left. The donor's intrathoracic IVC is anastomosed end to side to the external jugular vein of the recipient by a running stitch with a 10-0 nylon suture in a clockwise direction (Fig. 12.2). Herein, it is important to utilize ice-cold saline to flush donor's intrathoracic IVC to avoid kinking owing to its sufficient length of free part. To minimize bleeding, a cotton bud is applied around anastomotic sites before the microvascular clamps and the ties are removed. The proximal microclamp on the external jugular vein is first released, unclamped for the distal clip, and then the distal and proximal knots of the carotid artery are untied in this order to avoid anastomotic bleeding due to the formation of high pressure. The transplanted graft is immediately reperfused and then turned red in color. Warm water (30–37 °C) is dropped onto the graft to accelerate recovery of full beating with a sinus rhythm. At the end, the wound is continuously stitched with a 5-0 absorbable suture and the transplanted recipient is placed on a water-circulating heating pad until conscious.

One of the advantages of this cervical transplant model is the avoidance of emergence of tension by utilizing the donor intrathoracic IVC for vessel reconstruction, in which the ideal surgical area is exposed and favorable manipulation space is created. Importantly, the sufficient length of intrathoracic IVC is of great help in vascular anastomosis as the large distance exists between the carotid artery and the external jugular vein. A novel pattern of blood flow is formed in this model, in which blood

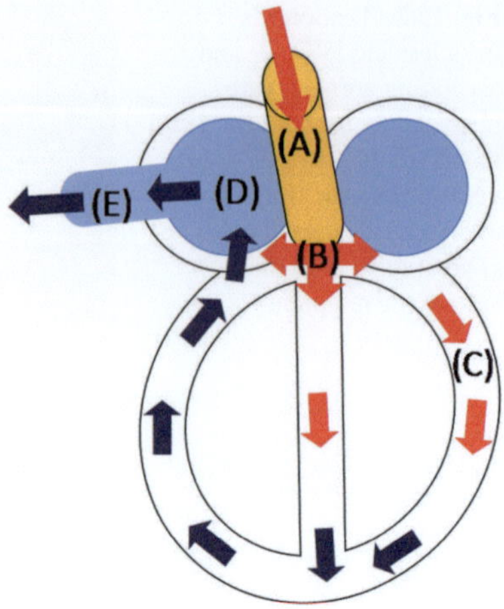

Fig. 12.3 This diagram is designed to describe the blood flow after vascular reconstruction in this transplant model. (*A*) The ascending aorta, (*B*) origin of coronary artery, (*C*) myocardium, (*D*) right atrium, and (*E*) the intrathoracic IVC

from the recipient's carotid artery enters donor coronary artery via the ascending aorta, passing through myocardium, draining into right atrium, and eventually flowing to the recipient's external jugular vein via the donor's intrathoracic IVC (Fig. 12.3).

Postoperative care, analgesia, graft assessment, and admission standard of transplanted recipient can be referenced by the description at the part of "Mouse abdominal heart transplant model". The advantage of this cervical cardiac transplant model in graft assessment is the possibility of direct observation of palpation.

12.4 Discussion

Since the first vascularized heterotopic heart transplant model was successfully established by Mann and associates in 1933 (Tomita et al. 1997), 80 years passed and a couple of modifications have been made to achieve a stable and fast model for scientific studies. In recent years, the direction of physiological blood flow of heart has been underscored although the cardiac graft de facto does not have any blood-pumping function.

This present model offers a several unique advantages: (1) compared with traditional cuff technique, this suture technique prevents anastomotic sites from thrombosis, instrumental in observing long-term graft survival; (2) replacement of donor pulmonary artery with donor intrathoracic IVC for vessel reconstruction makes more spaces and lessen the intensity for vascular anastomosis owing to sufficient length of donor intrathoracic IVC; and (3) the recipient's carotid artery and external jugular vein are preserved, which ensures recipient's blood supply of carotid artery to the central nervous system (Gong et al. 2012). As a result, the long-term graft survival was successfully achieved without any surgical complications.

However, to establish this transplant model, some key points arise for the beginners (1) owing to the limited space at cervical part, two miniclips are respectively suggested to leave at proximal end of external jugular vein and at the distal end of carotid artery. Two slip knots with 6-0 silk are respectively placed at the proximal end of carotid artery and at distal end of external jugular vein; (2) the placement of miniclip on the external jugular vein is also beneficial for better exposing the longitudinal venotomy; (3) accounting for the length of donor intrathoracic IVC, a kinking should be avoid with caution; (4) care should be taken to lease miniclips and slip knots in order (proximal end of external jugular vein, distal end of external jugular vein, distal end of carotid artery, and proximal end of carotid artery); and (5) during the implantation process, minced ice or cold saline should be given to protect graft (Gong et al. 2012).

12.5 Mouse Cervical Heart Transplant Model with Cuff Technique

Guangxiang Liu

Mouse heart transplantation is a method that is commonly used to investigate rejection mechanisms and immune response (Wang et al. 2008; Denecke et al. 2012). This technique is useful for immunological investigations of graft rejection, for detection of the therapeutic effects of immunosuppressive reagents, and for studies of the efficacy of new surgical techniques. The main advantages of mouse heart transplantation are that evidence of functional graft viability, and of heart beating, easily monitored by palpation. Furthermore, large quantities of monoclonal antibodies and material for molecular studies can be harvested for subsequent use. Therefore, the mouse model is clearly a valuable tool in many fields of medical research.

The technique most frequently used for heterotopic heart transplantation in mice was first described by Corry et al. in 1973 (Corry et al. 1973) and was later improved by many researchers (Tomita et al. 1997; Wang et al. 2005; Gong et al. 2012; Fang et al. 2013). This model is used as a routine procedure in some laboratories. However, the complexity of the surgical procedure has limited its widespread use of mouse vascularized heart transplantation as a medical research tool. Therefore, to simplify the surgical procedure and increase the postoperative survival rate, a cuff technique is developed for heterotopic cervical heart transplantations in mice. In our transplant model, the inferior and superior vena cava of donor are ligated and transected respectively. And the ascending aorta is isolated and transected at the level of the initial part of the innominate artery. The main pulmonary artery (PA) is transected at a little above the bifurcation of the PA, and then the residual blood vessels and lungs are ligated proximally and then carefully transected. The right external jugular vein and common carotid artery (CA) of the recipient are equipped with 20-gauge and 18-gauge cuffs, respectively. Using this cuff technique, the procedure from harvesting the donor heart to skin closure of the recipient mouse can be completed within 45 min, and the cold ischemic time for the graft is minimized to less than 20 min. The cervical heart transplantation with cuff technique in mice is a practical and reliable animal model and can be done by surgeons with a little microsurgery experience.

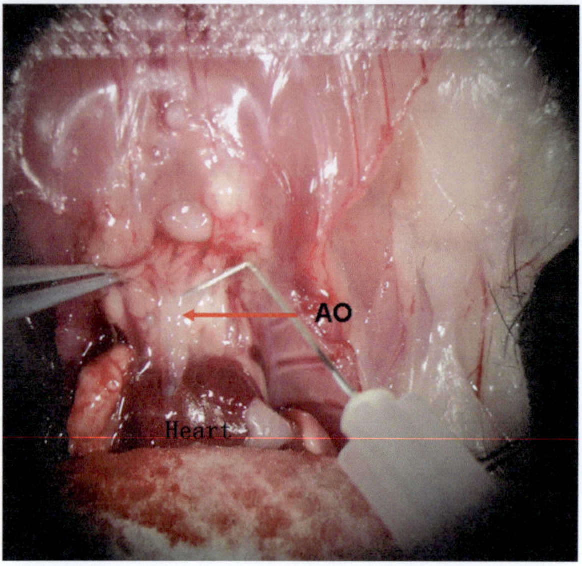

Fig. 12.4 The heart is perfused with 4 °C saline solution through the aortic arch

12.5.1 Materials and Methods

12.5.1.1 Animal

Male or female C57BL/6 (B6, H2b) and DBA/2 (H2d) mice weighing 22–28 g are used as donors and recipients. All animals are housed in an approved vivarium. Animals are allowed free access to food and water pre- and postoperatively. Donor and recipient animals are been anesthetized with either ketamine 80–100 mg/kg and xylazine 5–10 mg/kg or isoflurane (1–3 %).

12.5.1.2 Donor Procedure

Anesthesia has been induced by intraperitoneal injection of Xylazine and Ketamine. A long midline abdominal incision is made. Abdominal contents are gently retracted with gauze to left in order to expose the abdominal aorta and the inferior vena cava (IVC). Total anticoagulation will be induced by injecting 2,500 I.E. (0.5 ml Heparin (Heparin Immuno®), 5,000 I.E./ml) intravenously. The IVC and abdominal aorta are dissected and the animal has been bled to death. Next, the diaphragm has been dissected, the ribs have been cut bilaterally in the posterior axillary line, and the ventral thoracic wall has been shifted cranially. The IVC and superior vena cava (SVC) are isolated and dissected partly for outflowing of perfusate. The heart is perfused with 4 ml 4 °C saline solution through the aortic arch (Fig. 12.4). The IVC and SVC are ligated with 7-0 silk and transected. Blunt dissection of the aorta (AO) (Fig. 12.5) and pulmonary artery (PA) is performed with a little above the bifurcation of the PA preserved. The AO is retrogradely freed from the descending aorta, and the

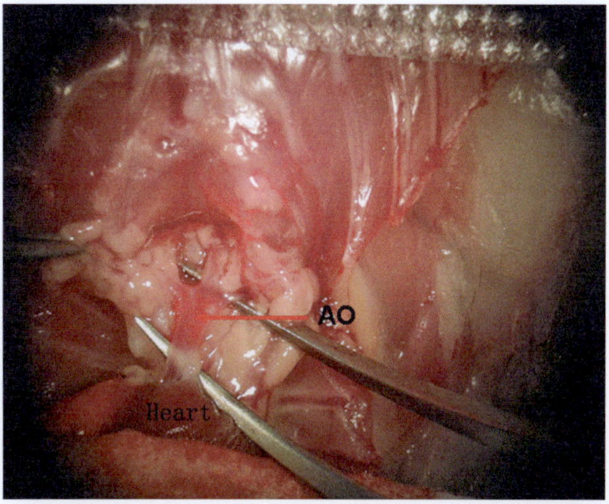

Fig. 12.5 Blunt dissection of the aorta

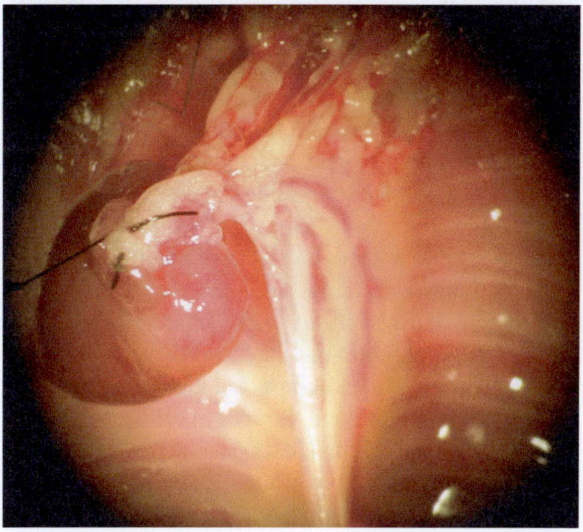

Fig. 12.6 All the vessels of the heart are ligated en bloc, except the AO and PA

ascending aorta is transected at the level of the initial part of the innominate artery. Finally, pulmonary veins and the left SVC are en bloc ligated (Fig. 12.6) and transected, the graft has then been removed from the thorax at once (Fig. 12.7). The heart is stored in ice-cold saline solution or transplanted immediately.

12.5.1.3 Recipient Procedure

The recipient is anesthetized as described previously. A longitudinal incision from the right mandibular angle to the middle point of the clavicle is made. After

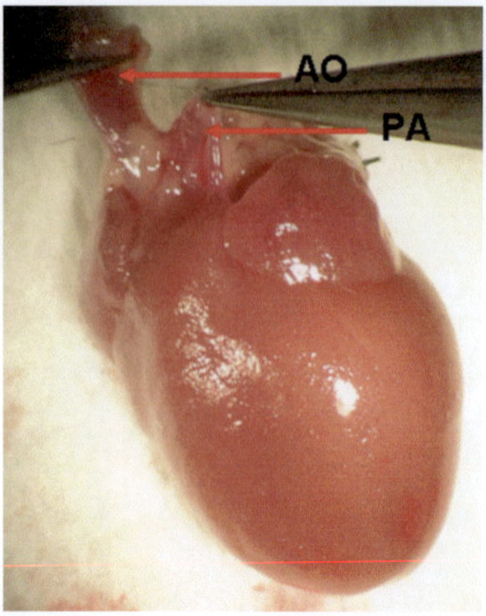

Fig. 12.7 The heart graft is harvested

resection of the right submandibular gland, the right external jugular vein is exposed carefully. The distal portion of the right external jugular vein is ligated (Fig. 12.8) at the level of the bifurcation with two 7-0 silks respectively and transected between two ligatures. The right external jugular vein is then passed through the venous cuff with a ligated suture on the free edge of it (Fig. 12.9). The proximal potion of the right external jugular vein is occluded with a small vessel clamp as proximally as possible (Fig. 12.10). By pulling the free edge of the right external jugular vein with forceps, the distal end of the right external jugular vein is everted over the cuff (Fig. 12.11) and ligated with a circumferential ligature of 7-0 silk (Fig. 12.12). The right common carotid artery (CA) is exposed by cutting the right sternocleidomastoid muscle and mobilized to the bifurcation of the internal and external carotid artery. The distal portion of the CA is ligated at the level of the bifurcation with two 7-0 silk respectively and transected the CA between two ligatures (Fig. 12.13). The CA is then passed through the artery cuff with a ligated suture on the free edge of it. The proximal potion the CA is occluded with a small vessel clip as proximally as possible (Fig. 12.14). By pulling the free edge of the CA with forceps, the distal end of the CA is everted over the cuff (Fig. 12.15) and ligated with a circumferential ligature of 7-0 silk (Fig. 12.16). As a next step, the donor heart is transferred to the neck of the recipient. The venous and arterial cuffs are inserted into the donor PA (Fig. 12.17) and AO (Fig. 12.18), respectively, and ligated circumferentially with 7-0 silks. During the 5 min of warm ischemia, the graft is been moistened at frequent intervals with cold saline solution. Finally, the venous and arterial clamps are removed (Fig. 12.19) and the transplant is reperfused immediately (Fig. 12.20). Upon rewarming and coronary perfusion, the heart began to fibrillate, and within a few minutes it reverted spontaneously to sinus rhythm. The incision is closed using 5-0 Prolene, nonabsorbable suture (Fig. 12.21). The mouse is then been exposed to 35 °C air temperature for recovery.

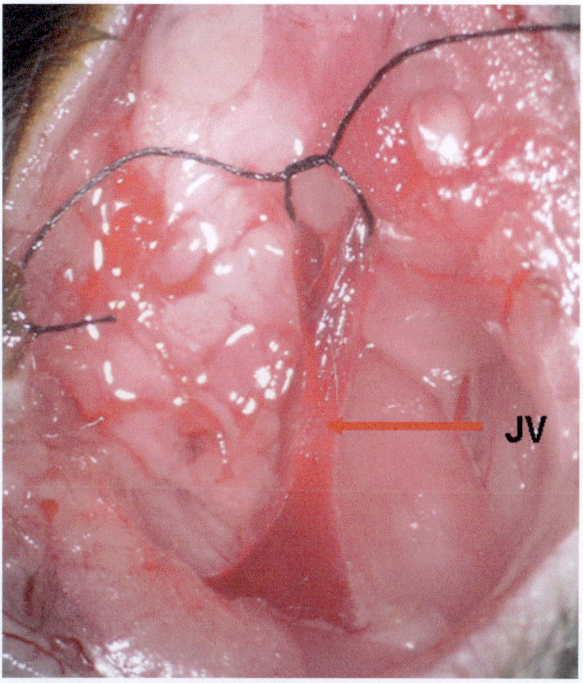

Fig. 12.8 To ligate the distal portion of the right external jugular vein

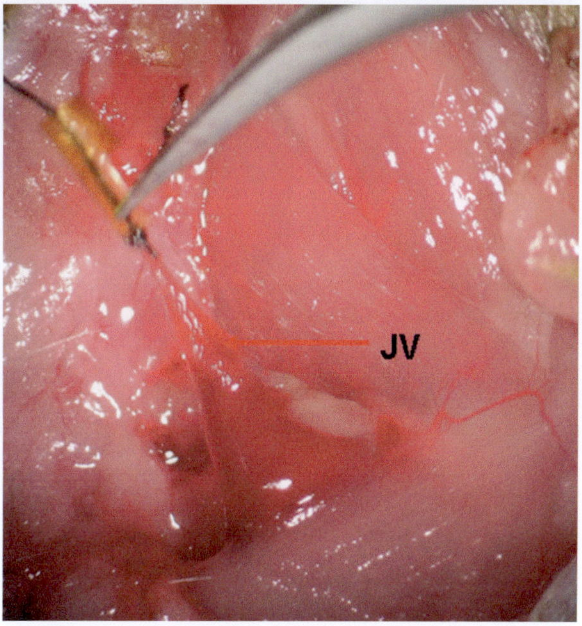

Fig. 12.9 The venous cuff is passed through the right external jugular vein

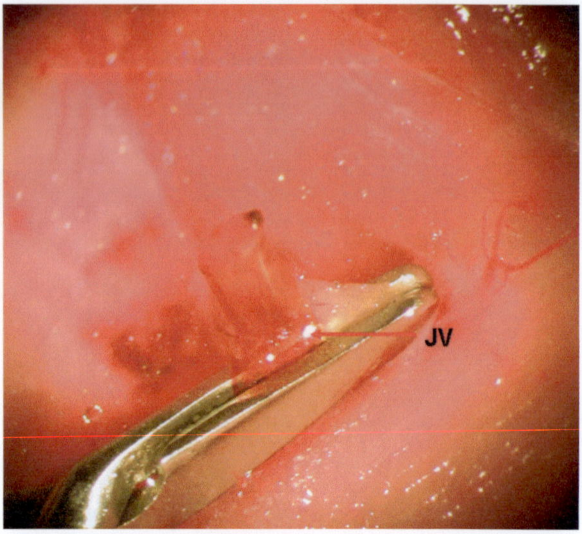

Fig. 12.10 The proximal portion of the right external jugular vein is clamped with a small vessel clip

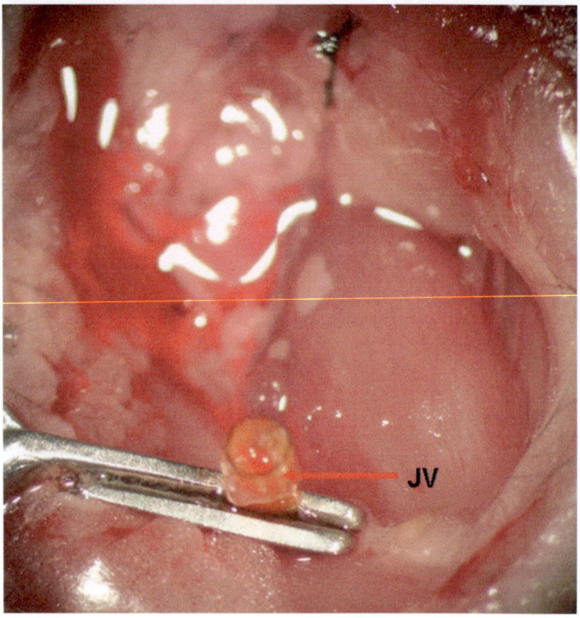

Fig. 12.11 The distal end of the right external jugular vein is everted over the cuff

12.5.2 Discussion

The cervical heterotopic heart transplantation model is outstanding, because of the superficial position of the donor heart in the neck, and therefore the heartbeat can be seen or be palpated. A suture technique for cervical heart transplantation in mice has not been as popular as abdominal heart transplantation for the technical

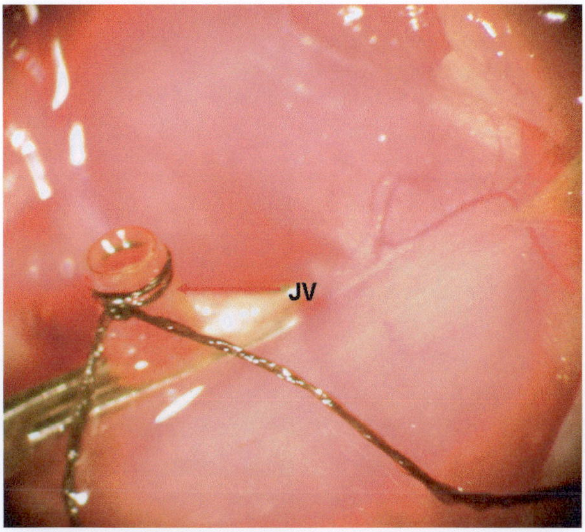

Fig. 12.12 The distal end of the right external jugular vein is ligated with a circumferential ligature

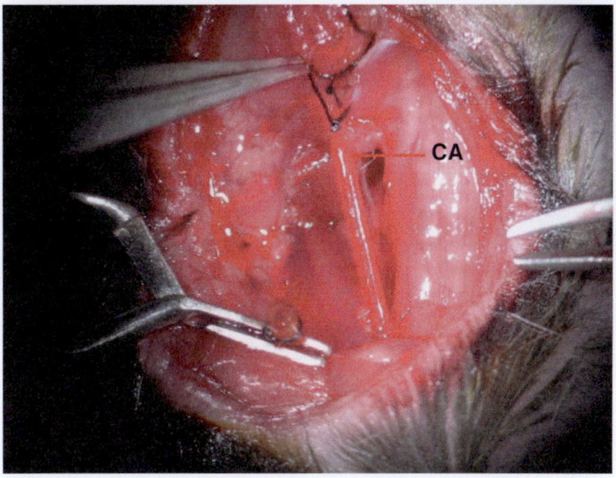

Fig. 12.13 To ligate the distal portion of the CA

difficulties in vessel anastomoses. In addition, the procedure is associated with two main complications of thrombosis and anastomosis leakage. Matsuura et al. (1991) then used a cuff without requirement of difficult vascular sutures, but three retention sutures remain needed. Tomita et al. (1997) used a special cuff with fixed arms. Though there is no need of retention sutures, the specially made cuff is difficult to acquire. Wang et al. (2005) prepared the cuff on PA of the graft. PA wall is very thin and easy to tear up. Moreover, it is sometimes impossible to make the cuff when the pulmonary artery trunk is too short. To improve the surgical procedures, we tried to fix the cuff on the recipient external jugular vein, which is proved to be easy. To reduce warm ischemic time, we perfused the heart with 4 °C saline solution through the aortic arch immediately after thoracotomy, which

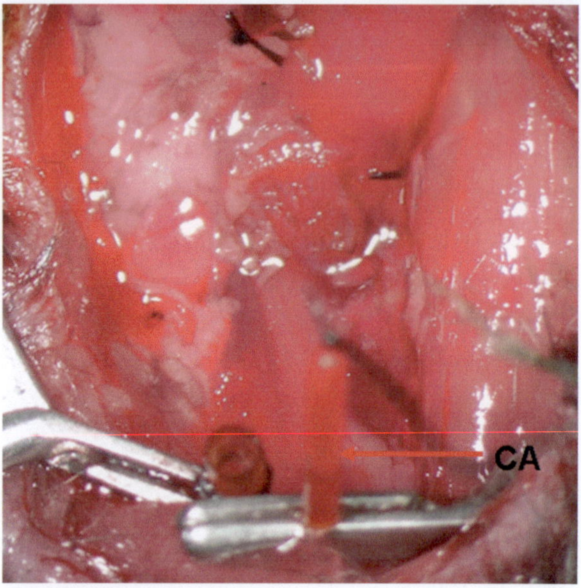

Fig. 12.14 To occlude the proximal potion the CA with a small vessel clamp

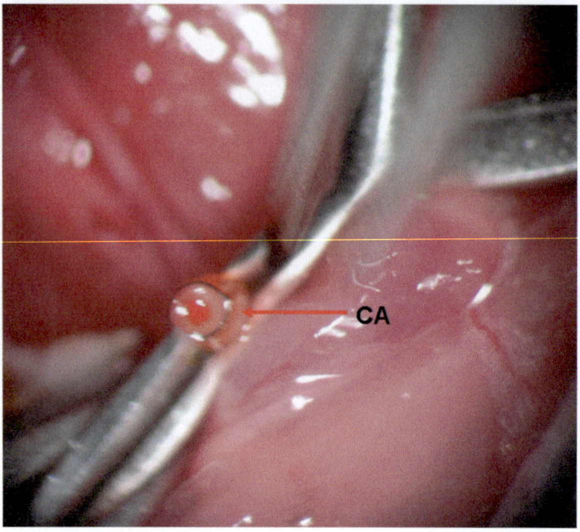

Fig. 12.15 The distal end of the CA is everted over the cuff

rapidly induced asystole in the donor heart to further reduce the effects of warm ischemia rendered donor heart became asystole rapidly. In addition, we paid attention to not twist the two vessels.

New researchers are easy to learn the cuff technique, because of no requirement of vascular anastomosis. We used 18G and 20G intravenous catheter Teflon tube as the arterial and venous cuffs, which are easy to obtain. The difficult part of cuff

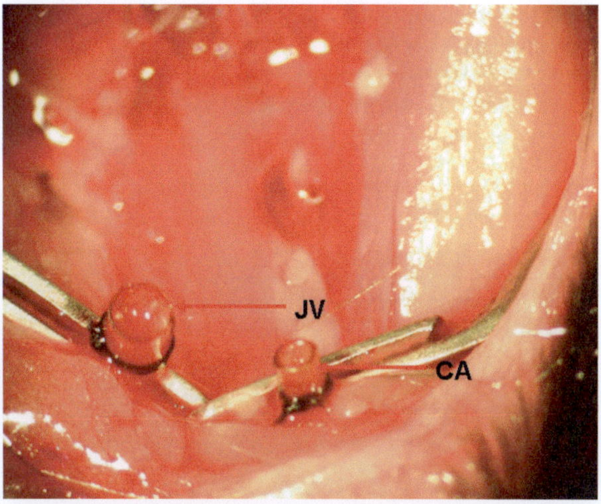

Fig. 12.16 The distal end of the CA is ligated with a circumferential ligature

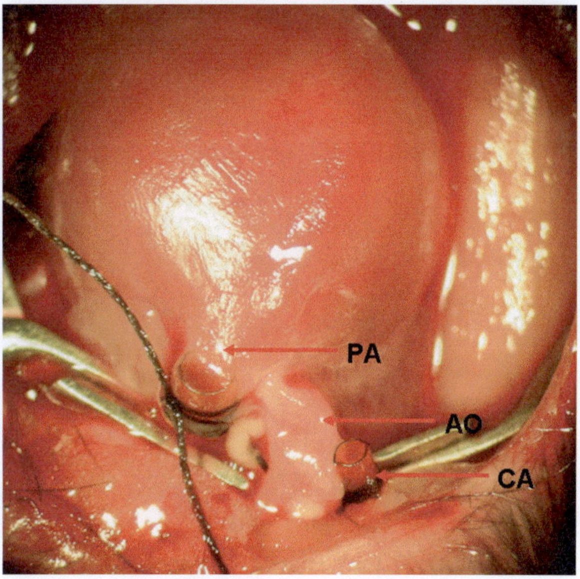

Fig. 12.17 The venous cuff is inserted into the donor PA and ligated circumferentially

technique is to evert the vessels over the cuffs. We cut at the bifurcation of recipient external jugular vein, as well as CA, which helped enlarge the end of the vessels, and thus eversion of the vessels became easy. The cuff procedure must be done gently and carefully.

In conclusion, this cuff technique offers a fast, convenient, and effective method for cervical heterotopic heart transplantation and, importantly, can be learned easily by beginners to the procedure.

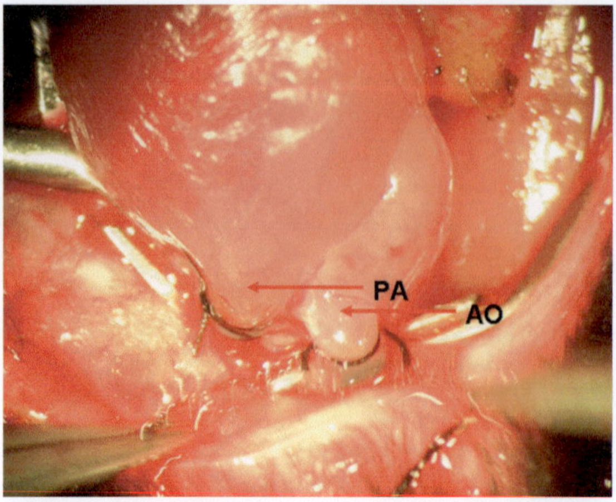

Fig. 12.18 The arterial cuff is inserted into donor AO and fixed circumferentially

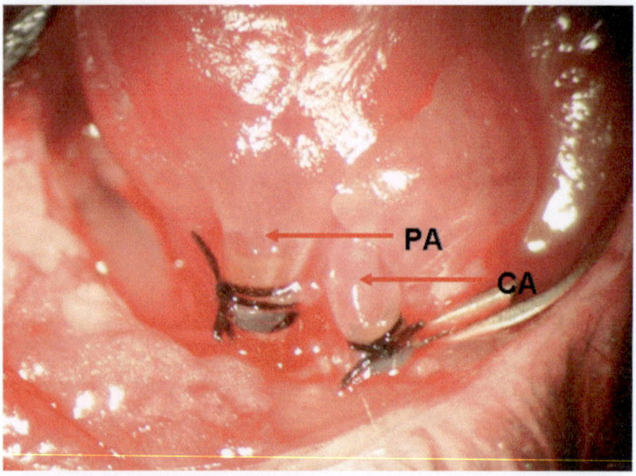

Fig. 12.19 The clamp on the right external jugular vein is unclamped, followed by unclamping of the CA

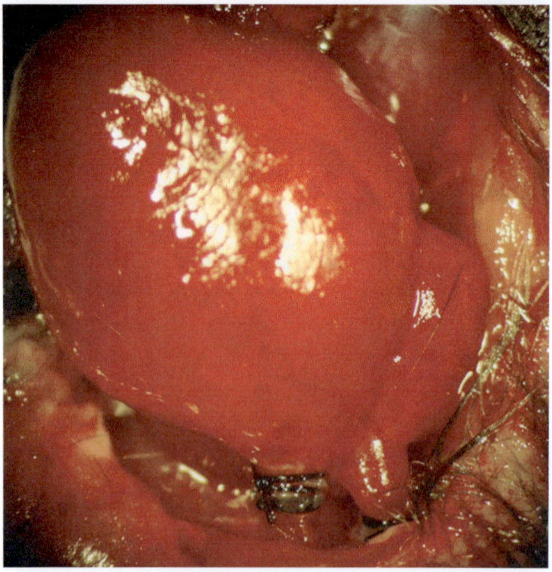

Fig. 12.20 The heart rapidly turned from pale to red

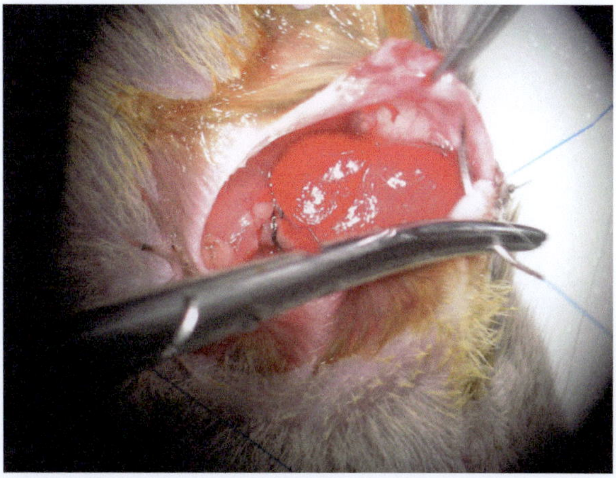

Fig. 12.21 To close the incision

References

Chen ZH. A technique of cervical heterotopic heart transplantation in mice. Transplantation. 1991;52(6):1099–101.

Corry RJ, Winn HJ, Russell PS. Heart transplantation in congenic strains of mice. Transplant Proc. 1973;5(1):733–5.

Denecke C, Ge X, Jurisch A, Kleffel S, Kim IK, Padera RF, Weiland A, Fiorina P, Pratschke J, Tullius SG. Modified CD4(+) T-cell response in recipients of old cardiac allografts. Transpl Int. 2012;25(3):328–36.

Fang J, He L, Wang SQ, Ma MJ, Liu HY, Zhu XH, Zhu P, Wei X, Wang CY. A simplified two-stitch sleeve technique for arterial anastomosis of cervical heterotopic cardiac transplantation in mice. Am J Transl Res. 2013;5(5):521–9.

Gong W, Thornley T, Whitcher GH, Ge F, Yuan S, Liu DJ, Balasubramanian S. Introduction of modified cervical cardiac transplant model in mice. Exp Clin Transplant. 2012;10(2):158–62.

Heron I. A technique for accessory cervical heart transplantation in rabbits and rats. Acta Pathol Microbiol Scand. 1971;A79(4):366–72.

Matsuura A, Abe T, Yasuura K. Simplified mouse cervical heart transplantation using a cuff technique. Transplantation. 1991;51(4):896–8.

Tomita Y, Zhang QW, Yoshikawa M, Uchida T, Nomoto K, Yasui H. Improved technique of heterotopic cervical heart transplantation in mice. Transplantation. 1997;64(11):1598–601.

Wang Q, Liu Y, Li XK. Simplified technique for heterotopic vascularized cervical heart transplantation in mice. Microsurgery. 2005;25(1):76–9.

Wang W, Carper K, Malone F, Latchman Y, Perkins J, Fu Y, Reyes J, Li W. PD-L1/PD-1 signal deficiency promotes allogeneic immune responses and accelerates heart allograft rejection. Transplantation. 2008;86(6):836–44.

Chapter 13
Mouse Kidney Transplantation

Weihua Gong

Experimental organ transplantation in rodents drives the development of novel insights into underlying mechanisms of ischemia/reperfusion injury and chronic/acute rejection. Mouse kidney transplantation plays an important role in both scientific research and translational medicine. Since Skoskiewicz M et al. firstly established the model of mouse kidney transplantation in 1973 (Skoskiewicz et al. 1973), many attempts have been made to modify the techniques (Ge and Gong 2011). However, the intractable manipulation on the urinary continuity has frustrated microsurgeon to establish this model and de facto hampered its wide use. Based on past experience on rat kidney transplant model, many researchers adopted bladder-to-bladder anastomosis as a direct end-to-end anastomosis is infeasible to achieve owing to the extremely tiny diameter of mouse ureter. On the other hand, blood is difficult to supply the lower segment of donor ureter and the bladder patch, leading to tissue necrosis. Herein, we describe the ureteral implantation technique for urethral reconstruction, which was initially developed by Han WR et al. in 1999 (Han et al. 1999).

The mouse kidney transplant model is characterized by its immune responses, in which relatively weak allogeneic response can be elicited even in the recipient grafted fully MHC-mismatched donor kidney. Unlike mouse cardiac transplant model, the kidney allograft (Balb/c → C57BL/6) can normally survive for long term without any treatment (Meng et al. 2008).

W. Gong, MD, PhD
Department of Surgery, The Second Affiliated Hospital of Zhejiang University School of Medicine, Hangzhou, People's Republic of China
e-mail: weihua.gong@gmail.com

W. Gong (ed.), *Rodent Transplant Medicine*,
DOI 10.1007/978-94-017-9472-5_13

13.1 Animals

Male inbred C57BL/6 (B6; H-2^b) and BALB/c (B/c; H-2^d) at the age of 8–12 weeks are normally used for transplant experiments. Mice are cared for at pathogen-free animal facilities under standard conditions and fed water ad libitum. All experiments in compliance with the institution's guidelines for experimental animals are approved by the local animal ethics committee.

13.2 Special Surgical Instruments

Cautery is utilized to cauterize lumbar branches before arteriotomy and venotomy. A 0.36-mm Microfine needle or an iris scissors are used for arteriotomy. A 30-gauge needle or an iris scissors may be prepared for venotomy. A bent 30-gauge needle is applied to flush vascular lumen. The use of circulating water blanket is necessary to avoid hyperthermia or severe burn, which is also helpful for mouse recovery from anesthesia.

13.3 Preparation of the Surgeon

Microsurgeon should wear lab coat or disposable full Tyvek suit, a mask, hair bonnet, gloves, and shoes cover. If circumstance permits, air shower is taken prior to entering animal area. Sterile gloves are suggested to use for animal surgeries.

13.4 Anesthesia

General anesthetics are normally given by inhalation or injection into peritoneal cavity. As special costly equipment is required for isoflurane and oxygen inhalation, instead, injection with ketamine (100 mg/kg) and xylazine (10 mg/kg) is adopted.

13.5 Donor Surgery

The donor mouse is anesthetized and the hair from the surgical area is clipped using a razor with a #40 clipper blade. Thereafter, lubricant ophthalmic ointment (Tearfair or Lacrilube) is applied to the eyes to prevent corneal dryness and avoid any irritation during the surgery. The mouse is placed on the surgical board and the clipped skin is aseptically prepared using Betadine and 70 % alcohol solution. The disinfecting process begins at the center of the proposed incision and then proceeds to the periphery. Overuse of 70 % alcohol should be avoided to prevent hypothermia.

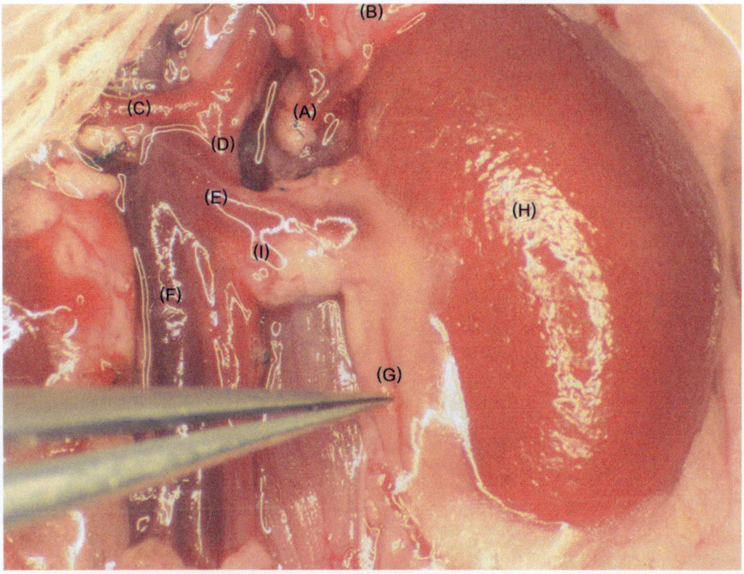

Fig. 13.1 (*A*) The free end of adrenal vein, (*B*) adrenal gland, (*C*) the superior mesenteric artery, (*D*) the suprarenal aorta, (*E*) left renal vein, (*F*) the infrarenal IVC, (*G*) the periureteral nourishing vessel, (*H*) left kidney, and (*I*) the free end of testicular vein

Like rat kidney transplant, left donor kidney is frequently used for transplant. After a midline incision from sternum to pubis is made, visceral organs are reflected to the right side of abdominal cavity to clearly expose the left kidney. The adrenal (A) and testicular (I) vessels are freed and tied with 9-0 sutures. The renal vessels (E), the suprarenal aorta (D) and IVC are dissected and the lumbar vessels are cauterized. A preset knot is prepared for the suprarenal aorta (A) and the infrarenal aorta with 10-0 nylon (C). The infrarenal aorta and IVC are tied with 6-0 thread (D) to prevent venous blood flow from the lower limbs during kidney perfusion. The superior mesenteric artery (B) is tied with 6-0 silk and transected to clearly expose the right renal artery and help prepare the segment of the suprarenal aorta for subsequent transplant anastomosis (Fig. 13.1). The left ureter is dissected from the renal hilum to the bladder. Care should be taken to protect the periureteral nourishing vessel to avoid ureter necrosis after surgery.

After the preplaced ligature of the suprarenal aorta (A) (Fig. 13.2) is tightly ligated to block blood flow, a short incision on the infrarenal IVC is made to drain subsequent perfusate and 0.2–0.4 ml heparinized chilled saline (50 U/ml) is immediately injected via the infrarenal aorta to perfuse kidneys with a low stable pressure. As a result, both kidneys uniformly become pale (Fig. 13.3). An elliptical excision is done at its junction with the IVC and an oblique incision between two renal arteries is made to transect the abdominal aorta. The left kidney graft is removed en bloc with urethral segment and placed in an ice-cold solution. The donor mouse is then sacrificed under anesthesia.

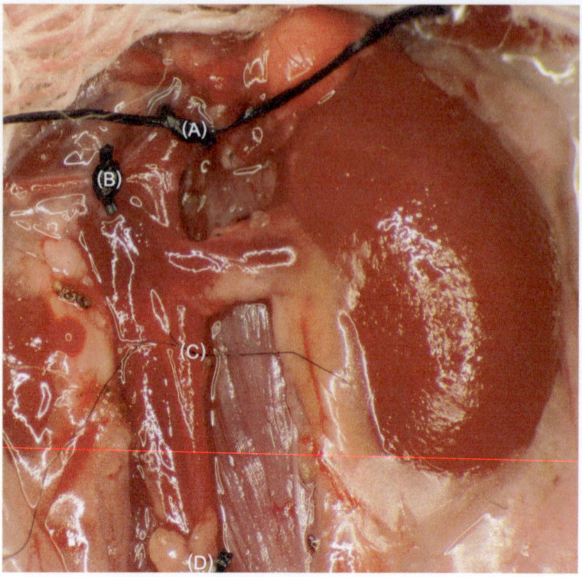

Fig. 13.2 (*A*) A preset knot around the abdominal aorta, (*B*) a tie of the superior mesenteric artery, (*C*) a preset knot around infrarenal aorta, and (*D*) a tie for the infrarenal aorta and IVC

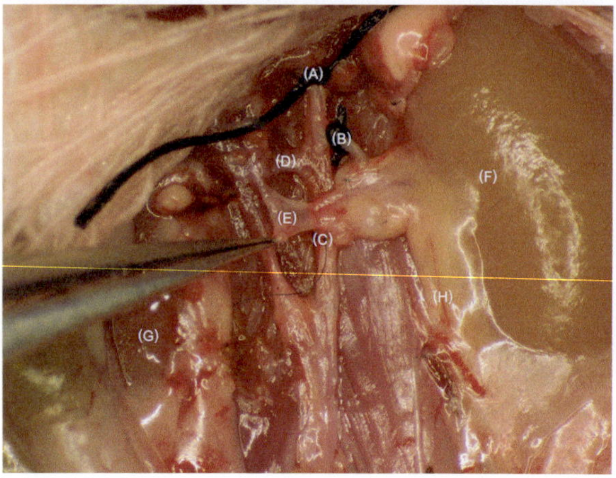

Fig. 13.3 (*A*) A tie of the suprarenal aorta, (*B*) the free end of the superior mesenteric artery, (*C*) a tie of the infrarenal aorta, (*D*) right renal artery, (*E*) left renal vein, (*F*) left kidney, (*G*) right kidney, and (*H*) left ureter

13.5.1 Cold Ischemia

Cold ischemia refers to the period starting from kidney perfusion, through procurement, and goes until the end of cold storage at 4 °C. Owing to highly demanding techniques of mouse model, only very few groups are capable of performing this procedure and therefore few data was generated about the deleterious effect of prolonged cold ischemia on graft injury or survival. It was shown that 2-h cold ischemia could cause a remarkable expression of intrarenal C3 protein, which is an acute-phase protein representing cold ischemia injury. When cold ischemic time is prolonged to 2.5–3 h, renal graft function will be impaired. And 4-h cold ischemia time will result in histological alteration such as severe renal tubular injury. Therefore, cold ischemia time should be controlled under 2.5 h.

13.5.2 Warm Ischemia

Warm ischemia means the period from positioning the graft in the recipient until blood flow reconstruction of graft. During the procedure of vascular reconstruction, microscopy lamplight and host's body temperature can increase the chilly graft's temperature, exacerbating the graft injury. Therefore, cold soaked saline gauge is suggested to wrap the kidney graft and sometimes ice is given to leave on the top of gauge to lower temperature. In general, warm ischemic time should be less than 35 min in mouse model, by contrast, which is shorter than that in rat renal transplantation.

13.6 Recipient Surgery

The mouse is anesthetized and the surgical area is prepared as described above. Laparotomy is performed and two self-made paper clip retractors are used to expose the abdominal contents. The intestine is mobilized onto the anterior wall of the chest and covered by wet gauze. Nephrectomy for left native kidney is performed by ligating left renal vessels and ureter with 6-0 silk. The lumbar veins are coagulated and the infrarenal aorta and IVC are freed from fat tissue. Two microserrefines (straight, 15-mm length, 100 g pressure, 6 × 1 mm jaw dimensions) are applied to cross-clamp the infrarenal aorta and IVC. A stitch with 10-0 nylon is placed through a full thickness of anterior median wall of aorta. An elliptical longitudinal aortotomy (B) is made by lifting 10-0 suture (Fig. 13.4). The length of opening is the same with or a bit smaller than the diameter of the free end of donor's aorta. A longitudinal venotomy (A) is directly made with an iris microscissors (Fig. 13.4). The donor kidney graft surrounding with ice-cold wet pad is then positioned to the right side of the recipient's abdominal cavity. Two stay sutures are placed at the distal and proximal apexes of the recipient's opening (Fig. 13.5) and then utilization of the proximal stay stitch continues to accomplish the artery reconstruction in a clockwise direction. As left side of arterial anastomosis is done, the kidney graft is reflected to the left flank of the recipient's abdominal cavity. On each side 3–4 running sutures are employed.

End-to-side venous anastomosis is constructed using a continuous 10-0 nylon suture (Ethicon, Inc.) (Fig. 13.6). Before the last knot is made to complete vein reconstruction, cold saline should be used to flush the anastomotic site to ensure patency and remove any intraluminal bubbles. Afterward, a bit of cotton pad is applied around arterial anastomotic site and the distal vascular clamp is firstly released. Venous blood flows into kidney graft. Subsequently, the proximal clip is immediately removed. On account of high blood pressure from the abdominal artery, an additional cotton swab is applied to press the arterial anastomotic site to

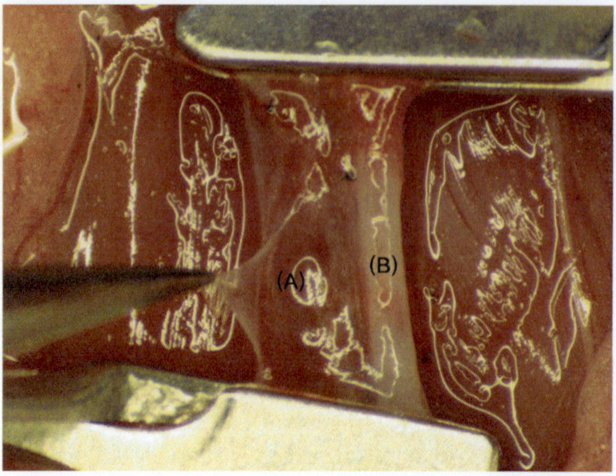

Fig. 13.4 An elliptical longitudinal aortotomy (*B*) is made by lifting 10-0 suture. A longitudinal venotomy (*A*) is directly made with an iris microscissors

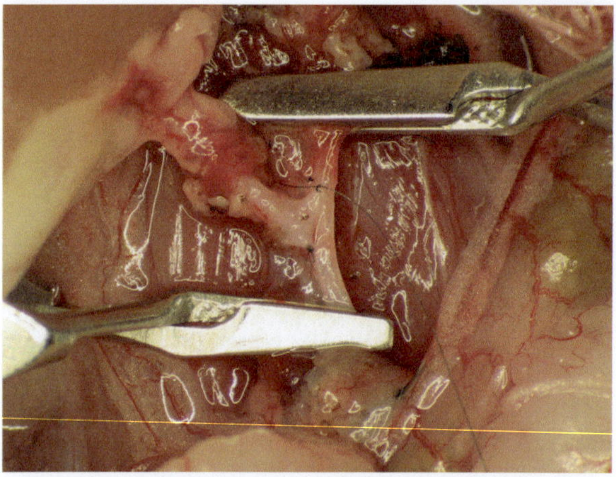

Fig. 13.5 Two stay sutures are placed at the distal and proximal apexes of the recipient's opening of the abdominal aorta

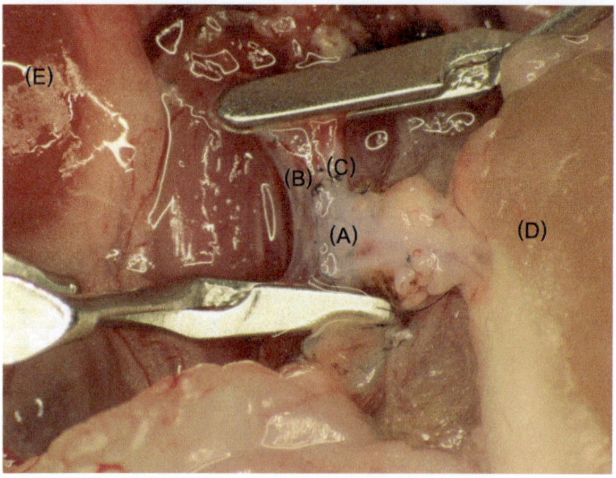

Fig. 13.6 (*A*) The donor's renal vein, (*B*) the recipient's infrarenal IVC, (*C*) the recipient's infrarenal aorta, (*D*) the kidney graft, and (*E*) the recipient's right kidney

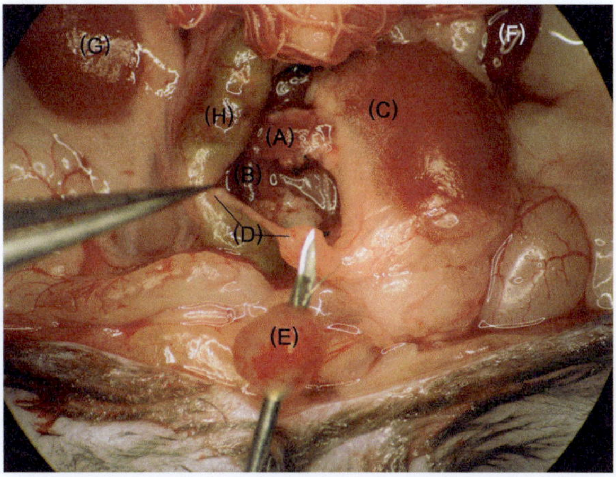

Fig. 13.7 (*A*) The donor's renal vein, (*B*) the recipient's IVC, (*C*) the reperfused kidney graft, (*D*) the donor's urethral segment, (*E*) the pierced bladder by 21-gauge needle, (*F*) the recipient's spleen, (*G*) the recipient's right native kidney, and (*H*) the recipient's large bowel

minimize bleeding. As no bleeding takes place, the cotton pad and swab are removed. The recipient's large bowel is moved to the right side of the graft.

13.6.1 Urethral Reconstruction

The ureter is reconstructed by urethral implantation technique. The distal segment (4–6 mm) of the donor's ureter (D) is further freed from its surrounding fat tissue and then a stitch with 10-0 suture and its needle is made at the endpoint of ureter as guidance. Herein, care should be taken to avoid damaging periureteral blood vessels. After this needle of 10-0 suture is straightened, the bladder (E) is diagonally punched from the low anterior wall to the high posterior wall by using a sterile 21-gauge needle and then 10-0 suture needle is inserted into the 21-guage needle cavity (Fig. 13.7). Subsequently, the 21-gauge needle is withdrawn together with 10-0 suture and its needle. The urethral segment can be pulled through both holes. Herein, one caveat is that care should be taken to prevent twisting of ureter by observation of periurethral vasculature. Afterward, the ureter is fixed to the bladder wall by interrupted suturing (2–3 stitches) (D) the periurethral tissue to the adventitia of the bladder around the posterior hole with 10-0 nylon (Fig. 13.8). An untied single purse-string suture with 10-0 nylon is made around the anterior hole. The distal end of the ureter is then incised to ensure the patency of urine flow and the "free segment" (2–3 mm) is left in the bladder cavity. Thereafter, the single purse-string suture is tied (Fig. 13.9).

The abdomen is closed by continuously stitching the two layers with 4-0 or 5-0 sterile, synthetic absorbable vicryl suture after the kidney graft is fixed by suturing its surrounding fat tissue to the recipient's retroperitoneum. 1.0 ml warm saline is subcutaneously injected to the recipient after surgery. Given much loss of blood, less than 0.5 ml normal saline could be additionally given through the penis vein.

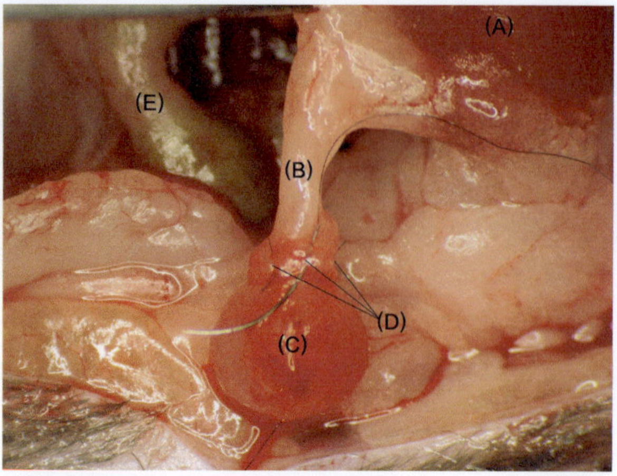

Fig. 13.8 (*A*) The reperfused kidney graft, (*B*) the donor's urethral segment, (*C*) the recipient's bladder, (*D*) three stitches for fixing the donor's urethral segment on the recipient's bladder, and (*E*) the recipient's large bowel

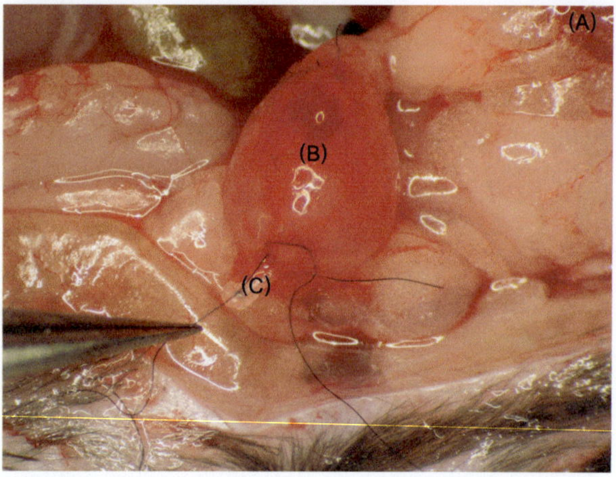

Fig. 13.9 (*A*) The reperfused kidney graft, (*B*) the recipient's bladder, and (*C*) purse-string suture to close the punch in the recipient's bladder

During the surgery, the following parameters should be recorded: cold ischemia time, warm ischemia time, and bladder reconstruction time.

13.7 Postoperative Care

13.7.1 Body Temperature

The smaller size of laboratory animals, the larger ratio of surface of area/body weight. Therefore, the mouse body temperature is apt to decrease in surgery. The circulating water blanket is normally used to increase and maintain body temperature for the first 24-h posttransplant. Noticeably, the mouse is housed in the cage to avoid hyperthermia after half of the cage is positioned on the heating blanket. Mouse usually recovers from anesthesia within 15 min after surgery and is given water, ad libitum and jelly gel.

13.7.2 Antibiotics

Although immunodefective mice are usually subject to transplant immunology studies, antibiotics are not mandatory to use for the sterile transplant surgery. However, a single intramuscular injection of Ceftriaxone or penicillin (500 U/10 g) may be given after surgery. Alternatively, sulfatrim (100 mg/kg) may be added in the drinking water.

13.7.3 Analgesic

Buprenorphine at the dose of 0.05 mg/kg is intramuscularly administered to the recipient for 48-h posttransplant surgery.

13.8 Nephrectomy for the Recipient's Native Kidney

Nephrectomy for the remaining native kidney in the recipient is performed at variable timepoints, which may take place at day 0, 4, 7, 10, and 14 posttransplant. Briefly, the mixture of ketamine (100 mg/kg) and xylazine (10 mg/kg) is firstly used to anesthetize the recipient. A midline incision is done and the intestine is gently moved to the right side of abdominal cavity. After the right kidney is clearly exposed, the right renal vessels and right ureter are ligated with 6-0 silk and the native kidney is removed. The abdomen is closed as described above.

13.9 Admission Standard of Transplanted Recipients

Bilaterally nephrectomized transplant recipient that does not survive more than 72 h should be considered a technical failure, regardless of cause.

13.10 Graft Assessment

Serum creatinine is normally measured using an I-Stat Portable Clinical Analyzer (Heska Corp., Fort Collins, CO) to reflect the graft function in the bilaterally nephrectomized recipient. Indeed, a mild hydronephrosis may occur in this transplant model owing to no antireflux structure for urethral reconstruction. Nonetheless, the serum creatinine will not increase without graft dysfunction, while the rejecting mouse will show signs of sickness apart from elevation of serum creatinine.

13.11 Histological Analysis

Paraffin-embedded or frozen graft sections are stained with hematoxylin and eosin (H&E) to evaluate the degree of lymphocytes infiltration, vasculitis, and tubulitis, which are morphological signs of a rejection process. The rejection can be quantitatively graded as: grade 1 = mild lymphocyte infiltration, (<25 % of section), no vasculitis, no tubulitis; grade 2 = moderate lymphocytes infiltration (25–50 % of section) with vasculitis and tubulitis; and grade 3 = severe lymphocytes infiltration (>50 % of section) with vasculitis and tubulitis. The grade should be scored in a blinded fashion. Noticeably, the structure of the different parts of kidney widely varies. Therefore, comprehensive analysis for graft pathology is required including renal cortex and medulla. The mean grade can be obtained by calculating all the grades at different locations. The formula can be found at the part of "Rat Kidney Transplantation".

The harvested kidney graft is fixed in buffered 10 % formalin and embedded in paraffin. The prevalence and degree of periodic acid Schiff (PAS) positivity are blindly evaluated for the stained sections according to the updated Banff classification.

13.12 Discussion

Although rat kidney transplant model has been used to advance our understanding of many clinically relevant transplant issues, mouse models remain preferred owing to their advantages in developing inbred laboratory lineage and ease of in vivo or in vitro manipulative reagents. Nevertheless, mouse renal transplant model has no overwhelmingly advantage as long-term allograft survival is spontaneously achieved without any treatment, i.e., spontaneous acceptance of graft. This probably results from a lower expression level of MHC molecules on the surface of endothelial cells and downregulation of T-cell receptors on the surface of CD8+ T cells in mice (Bedi et al. 2010).

Albeit mouse kidney transplantation has been successfully established for three decades, the goal of wide use of this model remains unaccomplished. A variety of challenging factors have been carefully reviewed by our group, including perfusion and preservation solution, technical notes, vascular reconstruction, urinary tract reconstruction, and operational complications (Ge and Gong 2011). Based on experienced microsurgical techniques from other rodent transplant models, major concern of mouse kidney transplantation is vascular anastomosis and urinary tract reconstruction. This chapter presents relatively simple, fast, and stable techniques for successfully establishing the model. Nevertheless, care should be taken to minimize warm/cold ischemia time and to avoid operational complications such as blood loss, thrombosis, urine leakage, etc.

References

Bedi DS, Riella LV, Tullius SG, Chandraker A. Animal models of chronic allograft injury: contributions and limitations to understanding the mechanism of long-term graft dysfunction. Transplantation. 2010;90(9):935–44.

Ge F, Gong W. Strategies for successfully establishing a kidney transplant in a mouse model. Exp Clin Transplant. 2011;9(5):287–94.

Han WR, Murray-Segal LJ, Mottram PL. Modified technique for kidney transplantation in mice. Microsurgery. 1999;19(6):272–4.

Meng L, Wu Z, Wang Y, Lassman C, Busuttil RW, Zhai Y, Kupiec-Weglinski JW. Differential impact of CD154 costimulation blockade on alloreactive effector and regulatory T cells in murine renal transplant recipients. Transplantation. 2008;85(9):1332–8.

Skoskiewicz M, Chase C, Winn HJ, Russell PS. Kidney transplants between mice of graded immunogenetic diversity. Transplant Proc. 1973;5(1):721–5.

Chapter 14
Mouse Liver Transplantation

Ulrich Steger, C. Otto, and I. Klein

Orthotopic mouse liver transplantation (MLTx) is an established model for immunological studies of graft rejection and tolerance induction after solid organ transplantation. Since its introduction in mice by Qian et al. (1991) in 1991, the mouse model has been frequently used to study immunological responses (Qian et al. 1994, 1997, 1996; Li et al. 2001; Klein and Crispe 2006). The model is also suitable to analyze the pathways of hepatic regeneration after prolonged ischemia time and/or partial liver transplantation (Conzelmann et al. 2003; Tian et al. 2011).

Liver allografts are less aggressively rejected than heart, kidney, skin, and intestinal grafts. In the mouse model, spontaneous acceptance of liver allografts occurs even in completely MHC-mismatched donor and recipient strain combinations (Table 14.1). Nevertheless, there is also a report by Sugioka et al. about reduced graft acceptance after mouse liver transplantation using wild-type mice with a high disparity in the MHC (Sugioka et al. 2001). The mechanisms leading to graft acceptance and induction of alloantigen-specific tolerance or unresponsiveness without immunosuppressive agents are still not fully understood. Investigations of the immunological mechanism of liver tolerance in mice may lead to improved therapeutical regimes for rejection after clinical kidney, heart, pancreas, or liver transplantation.

The technique of mouse liver transplantation is adapted from the well-established rat liver model to maximize the advantages of more knockout models, lower antibody consumption, and much greater availability of analytical tools. However, the model requires a much higher microsurgical skill level due to the smaller vessel size and the greater fragility of mouse liver tissue upon surgical manipulation and depression of the cardiovascular system by the anhepatic phase and overall procedure time.

U. Steger, MD (✉) • C. Otto, Ph.D • I. Klein, MD
Department of Surgery I, University of Wuerzburg, Wuerzburg, Germany
e-mail: klausulrich.steger@googlemail.com

W. Gong (ed.), *Rodent Transplant Medicine*,
DOI 10.1007/978-94-017-9472-5_14

Table 14.1 Survival and graft acceptance after allogeneic mouse liver transplantation in different strain combinations

Strain combination	Median survival (days)	Graft acceptance	Author
Major mismatch (I+II+minor)			
BL10→C3H	>100	83 %	Qian et al. (1994)
BALB/c→C3H	16	33 %	
C3H→BL10	35	20 %	
BL6→C3H	>100	73 %	Zhang et al. (1996)
BALB/c→CBA	>100	57 %	
BL6→BALB/c	>100	72 %	
BL10→CBA	>100	95 % (n=20)	Steger et al. (2003)
CBA→BL10	>100	100 % (n=4)	Steger U (2002) (unpublished data)
MHC I mismatch			
CBK→CBA	>100	75 % (n=4)	
Minor mismatch			
BL10.BR→CBA	>100	100 % (n=2)	

14.1 Surgical Procedure

Due to the anhepatic time with less circulation and consecutively less anesthetic demand, the recipient operation is performed under inhalation anesthesia. This provides well-regulated anesthesia during the operation with a rapid postoperative recovery. Methoxyflurane or Isoflurane are strong analgesic nonhepatotoxic volatile agents providing a good muscle relaxation. Isoflurane is supplied by means of an anesthetic machine including a vaporizer, mouse face mask, and scavenger system.

14.2 Donor

The donor operation can be performed under injection of a high dose mixture of, e.g., Fentanyl/Fluanisone (Hypnorm™) and Midazolam (Hypnovel™). After antiseptic wash of the abdominal area, an abdominal midline incision is made and extended to the lateral abdomen below the thorax on both sides. The intestines are rotated out of the abdominal cavity and wrapped in moistured gauge. First, the liver is freed from its ligaments to the diaphragm and stomach followed by a cholecystectomy. Then, the right suprarenal vessels and the pyloric vein are ligated and transected.

In the arterialized liver transplantation model, the hepatic artery is dissected toward the abdominal aorta by ligating the side branches of the celiac axis. Then, the left renal artery and vein, and all the small arteries rising from the abdominal aorta from the bifurcation upward are ligated and divided, except for the right renal, superior mesenteric, and coeliac arteries.

Next, the common bile duct is identified and ligated close to the pancreas. A 25-G catheter (polyethylene tube, of about 4 mm length, outer diameter ca.

0.5 mm, e.g., by Hell+Co. GmbH, Diespeck, Germany or Pajunk GmbH, PA-Catheter, Geisingen, Germany) is inserted toward the liver and secured by a circular tie (8-0 silk) (Fig. 14.1). An overview of various intraluminal stent tubes used for reconstruction of the bile duct during mouse liver transplantation is given by Chen et al. (Chen et al. 2013). After double clamping and dissection of the inferior vena cava infrahepatically, the liver is perfused with 3–5 ml cold saline solution given through the portal vein by a 24G i.v. catheter. The suprahepatic vena cava is cut near the diaphragm. Finally, the liver is carefully removed and stored in a 4 °C cold saline bath. Heparin is not used.

In the arterialized model, the right renal artery, the superior mesenteric artery, and the abdominal aorta proximal of the truncus coeliacus are ligated after clamping the infrahepatic vena cava and directly before starting with the cold perfusion of the liver.

14.3 Back Table

In the bath of cold saline, a 2-mm cuff (polyethylene tube, e.g., Braun Vasofix 20G, 1.1 mm) secured around the vessel stump with a silk ligation facilitates the following anastomosis of the portal vein (Fig. 14.1). There are several descriptions of MLTx that reported the use of a second cuff for the infrahepatic vein reconstruction. After reperfusion the portal vein, the anhepatic time is over and consequently, the hemodynamic situation improves except for the venous drainage of the kidneys. In

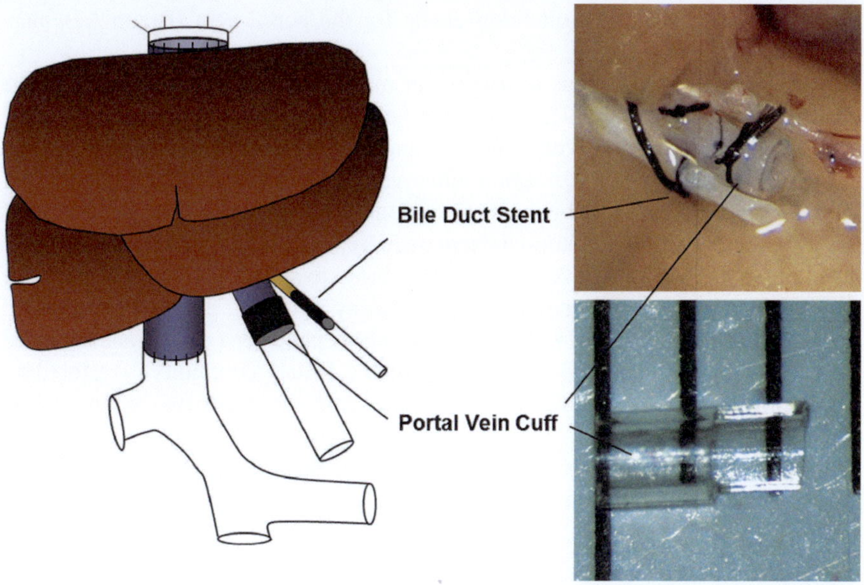

Fig. 14.1 Donor preparation for MLTx including the insertion of a portal cuff and biliary stent

our opinion, it seems easy to perform the infrahepatic cava anastomosis by a hand sewing technique. Humar et al. presented an MLTx model where all the vascular anastomoses were sewed without the use of any cuff (Humar et al. 2013). Hand sewing of the portal vein is time-consuming and requires a lot of microsurgical expertise. Even in trained hands, the anhepatic time will rise to a dangerous limit of around 25 min as mentioned by Humar et al. Even in long-term follow-up, we never experienced any complications using a cuff technique for the portal venous reconstruction and therefore appreciated the shorter anhepatic phase compared to the hand-sewn technique.

14.4 Recipient

The recipient operation is performed under inhalation anesthesia, described above, with concentrations of Isoflurane between 1.2 and 1.5 % to maintain deep and sufficient narcosis and analgesia. After antiseptic wash of the abdominal area, the abdomen is approached through an abdominal midline incision. The intestines are rotated out and wrapped in moist plastic foil to reduce the loss of fluid. After separating the surrounding ligaments, the right adrenal vessels, pyloric vein, hepatic artery, and common bile duct are ligated and transected.

The infrahepatic vena cava and the portal vein are then cross-clamped successively (start of anhepatic time). The liver is gently pulled down and the suprahepatic vena cava cross-clamped with a vascular clamp. There are several types of clamps that can be used. We prefer a curved temporary Yasargil Titan Miniclip for the subdiaphragmatic/suprahepatic vena cava, a straight Yasargil Miniclip at the infrahepatic vena cava and a microbulldog clamp for the portal vein (Braun-Aesculap, Tuttlingen, Germany).

After transecting the veins near the liver margins, the liver is removed and the donor graft placed orthotopically into the field. First, the suprahepatic vena cava anastomosis is constructed with a 360° running suture using 10-0 Nylon (Fig. 14.2a). Then, the portal vein cuff is inserted into a venotomy made in the anterior wall of the recipient portal vein that is kept on tension with a mosquito clamp at the tip of the venous stump. The venotomy is kept open by continuously flushing saline into the stump during insertion. A silk suture is placed to secure the vein over the cuff and the liver is revascularized after 12–17 min by removing the suprahepatic and portal venous clamps (Fig. 14.2b). Next, the infrahepatic vena cava is anastomosed by a continuous suture (10-0 Nylon) (Fig. 14.2c). Biliary continuity is restored by connecting the ducts over the donor stent and securing them with two or three single stitches (10-0 Nylon).

In the rearterialization model, the donor abdominal aorta stump is anastomosed end to side to the recipient infrarenal aorta with a running suture (10-0 Nylon) from both sides (Fig. 14.3).

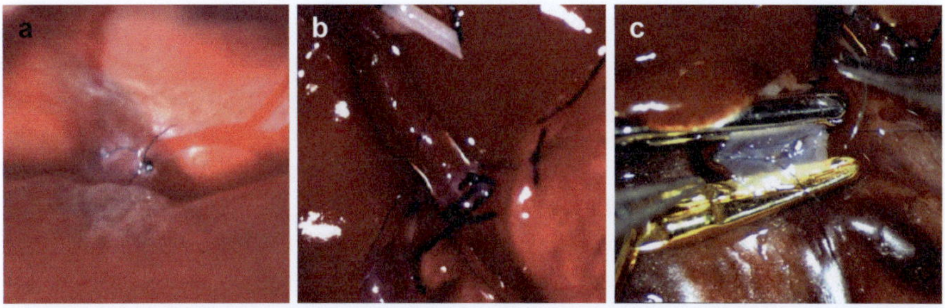

Fig. 14.2 Liver implantation with (**a**) a hand-sewn suprahepatic vein anastomosis, (**b**) cuff insertion at the portal vein, and (**c**) running suture at the infrahepatic cava anastomosis

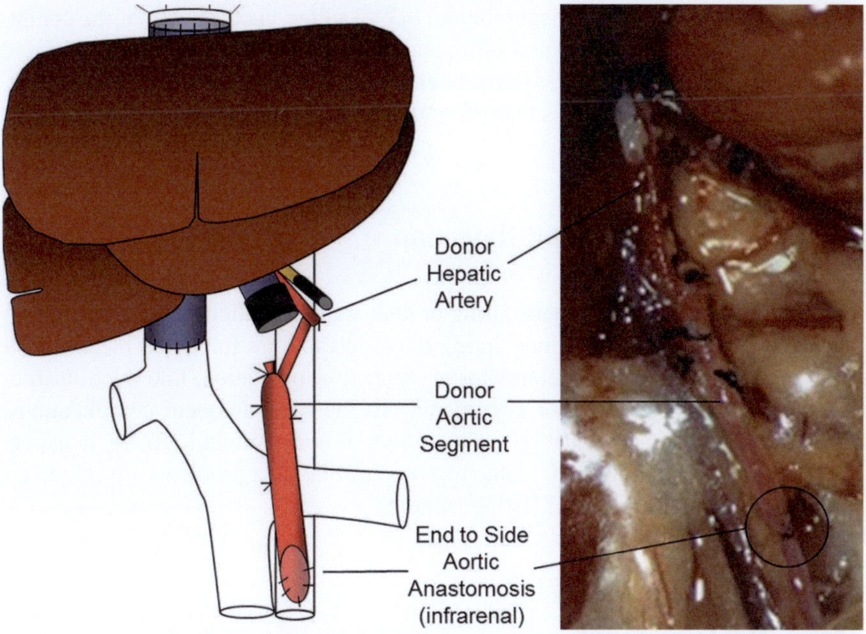

Fig. 14.3 Rearterialization of the hepatic graft by an end to side anastomosis of a donor aortic segment on the abdominal aorta of the recipient

Blood and fluid lost during operation is replaced during the procedure by intraperitoneal and intravenous injection of normal saline solution. After injection of local anesthetics around the incision margins, the abdominal cavity is finally closed in two layers with absorbable sutures. Postoperative, a single subcutaneous injection of 1 mg/kg Buprenorphine in 1 ml saline is given. The mice are kept in a warm climate chamber for the first 24 h. The recipients received neither heparin nor perioperative antibiotics.

14.5 Tips and Pitfalls

- To improve the clamping maneuver of the suprahepatic vein at the diaphragm, place a silk thread around the vein and pull down the liver with the thread.
- Sew the suprahepatic vein from a cranial position by turning the mice 180° with the nose pointing toward the surgeon.
- Flush the suprahepatic vena cava before sewing the last stitch at the suprahepatic anastomosis to avoid air emboli.
- The anhepatic time should be less than 20 min.
- Reduce the concentration of the inhalation anesthesia (e.g., isoflurane) during the anhepatic time to around 50 % of the regular concentration.
- Give a supportive injection of intravenous fluid after completion of the vascular reconstruction in the recipient of about 0.3 ml saline via the angle of the confluence of the left renal vein and infrarenal vena cava or, in case of a arterialized model, via the donor aortic segment before suturing of the arterial anastomosis.
- Avoid young mice with less than 25 g body weight as recipients.

14.6 Complications and Rejection

Complications or adverse events during or after transplantation occur in about 5 % of the procedures. Complications during the procedure are fulminant bleeding, air embolism, thrombosis, or pneumothorax. After transplantation, bile duct obstruction or leakage, peritonitis, or pneumonia are the most frequent complications. Signs of an acute severe rejection are rapid loss of body weight (>20 %), jaundice, and general signs of distress. Chronic rejection or bile duct problems are characterized by a slow development of lasting jaundice.

14.7 Arterialization

In the rat liver transplantation model, the necessity of rearterialization is still under controversial discussion (see chapter of rat liver transplantation). The same discrepancy can be recognized in the mouse model. There are different results in the literature about the overall survival after syngeneic mouse liver transplantation performed with or without reconstruction of the hepatic arterialization (Table 14.2). By looking at the results in Table 14.1 in the allogeneic (all groups nonarterialized) and Table 14.2 in the syngeneic setting, one could argue that the value of arterialization is strain dependent. BALB/c mice seem to be more susceptible to the lack of arterial blood supply than other strains of mice such as CBA or BL6. The results obtained in the nonarterialized model may be different from the physiological situation with an intact arterial blood supply, as routinely carried out in human clinical liver

Table 14.2 Overall survival after nonarterialized and arterialized syngeneic mouse liver transplantation

Strain	Arterialized Y/N	Overall survival (%)	*N*	Observation time	Author
B6AF1	N	74	27	14 weeks	Qian et al. (1991)
C57BL/6	N	64	11		
BALB/c	N	50	10		
CBA	N	100	8	100 days	Steger et al. (2003)
CBA	Y	100	6		
BALB/c	N	50	8	14 days	Tian et al. (2002)
BALB/c	Y	100	8		

transplantation. Nevertheless, in CBA mice, histological and blood analyses of syngeneic recipients without rearterialization showed no significant short and long-term damages, especially of the biliary system (Steger et al. 2003).

The necessity of arterialization in the mouse liver transplantation model is still not clear. It seems to depend on the experimental setting, including the strain combination.

References

Chen J, Gong W, Ge F, Huang T, Wu D, Liang T. A review of various techniques of mouse liver transplantation. Transplant Proc. 2013;45(6):2517–21.

Conzelmann LO, Zhong Z, Bunzendahl H, Wheeler MD, Lemasters JJ. Reduced-size liver transplantation in the mouse. Transplantation. 2003;76(3):496–501.

Humar B, Raptis DA, Weber A, Graf R, Clavien PA, Tian Y. Sewed revascularization for arterialized liver transplantation in mice. J Surg Res. 2013;184(2):e1–7.

Klein I, Crispe IN. Complete differentiation of CD8+ T cells activated locally within the transplanted liver. J Exp Med. 2006;203(2):437–47.

Li W, Lu L, Wang Z, Wang L, Fung JJ, Thomson AW, Qian S. Il-12 antagonism enhances apoptotic death of T cells within hepatic allografts from Flt3 ligand-treated donors and promotes graft acceptance. J Immunol. 2001;166(9):5619–28.

Qian S, Demetris AJ, Murase N, Rao AS, Fung JJ, Starzl TE. Murine liver allograft transplantation: tolerance and donor cell chimerism. Hepatology. 1994;19(4):916–24.

Qian S, Fu F, Li Y, Lu L, Rao AS, Starzl TE, Thomson AW, Fung JJ. Impact of donor MHC class I or class II antigen deficiency on first- and second-set rejection of mouse heart or liver allografts. Immunology. 1996;88(1):124–9.

Qian S, Lu L, Fu F, Li Y, Li W, Starzl TE, Fung JJ, Thomson AW. Apoptosis within spontaneously accepted mouse liver allografts: evidence for deletion of cytotoxic T cells and implications for tolerance induction. J Immunol. 1997;158(10):4654–61.

Qian SG, Fung JJ, Demetris AV, Ildstad ST, Starzl TE. Orthotopic liver transplantation in the mouse. Transplantation. 1991;52(3):562–4.

Steger U, Sawitzki B, Gassel AM, Gassel HJ, Wood KJ. Impact of hepatic rearterialization on reperfusion injury and outcome after mouse liver transplantation. Transplantation. 2003;76(2):327–32.

Sugioka A, Morita M, Fujita J, Hasumi A, Shiroishi T. Graft acceptance and tolerance induction in mouse liver transplantation using wild mice. Transplant Proc. 2001;33(1–2):137–9.

Tian Y, Graf R, El-Badry AM, Lesurtel M, Furrer K, Moritz W, Clavien PA. Activation of serotonin receptor-2B rescues small-for-size liver graft failure in mice. Hepatology. 2011;53(1):253–62.

Tian Y, Rudiger HA, Jochum W, Clavien PA. Comparison of arterialized and nonarterialized orthotopic liver transplantation in mice: prowess or relevant model? Transplantation. 2002;74(9):1242–6.

Zhang Z, Zhu L, Quan D, Garcia B, Ozcay N, Duff J, Stiller C, Lazarovits A, Grant D, Zhong R. Pattern of liver, kidney, heart, and intestine allograft rejection in different mouse strain combinations. Transplantation. 1996;62(9):1267–72.

Appendix

In the animal studies, rats and mice will be our "patients". Therefore, it is great importance in recording clinical observations, including the time to harvest, the time to transplant the graft, cold ischemia time, warm ischemia time, vasospasm (defined as full perfusion flow through the graft after an initial period <10 min of slow reperfusion), recovery of grafts (urine production), and presence of complications for each group.

Record of Experiments

OP date				
Animal. No.		Perfusion solution		
Anesthetics				
		Heparin (IU)		
Complications		Ischemia time (min)	Cold	
			Warm	

	Weight (gram)	Gender	Genotype
Donor			
Recipient			

W. Gong (ed.), *Rodent Transplant Medicine*,
DOI 10.1007/978-94-017-9472-5

Treatment

Mice	No.	Weight (gram)	Day	Dose

Animal Surgical Record

Date	Animal NO	Animal strain	Operation	Anesthesia	Analgesia			Antibiotics			Drugs			Surgical outcome	Clinical observation		Surgeon
					Time	Dose	Initial	Time	Dose	Initial	Time	Dose	Initial		Day 1	Day 2	

Note: A. *Type of anesthesia*: (1) *isoflurane* inhalation; (2) *pentobarbital* 50 mg (rat)-60 mg (mouse)/kg I.P
B. *Analgesia*: (1) *carprofen* 6 mg/kg 0.03 ml/each (s.c.) q.d for 48 h; (2) *buprenorphine* 0.05 mg/kg b.i.d s.c for 48 h
C. *Antibiotics*: (1) *amoxil* 50 mg/ml oral daily for 7 days; (2) *mefoxin* 40 mg/kg b.i.d im
D. *Experiment drugs*:
E. *Operation*: (1) *O.L.T* orthotopic liver transplantation; (2) *H.H.Tx* heterotopic heart transplantation; (3) *Skin Tx* skin transplantation; (4) *K.Tx* kidney transplantation; (5) *S.B.Tx* small bowel transplantation
F. *Surgical outcome*: (1) alive; (2) normal and active; (3) weak; (4) dead
G. *Observations*: (1) heart and respiration rats; (2) consciousness and reflexes; (3) cyanosis; (4) dehydration; (5) temperature in the water circulating heat pad; (6) wound bleeding; (7) breathing: ((a) normal (b) rapid (c) very slow (d) labored); (8) vocalization; (9) sternal recumbency; (10) move around; (11) paralysis; (12) ascites; (13) normal active

Daily Clinic Observation

Title: **Protocol**:

Animal NO	Operation date	Operation	Time points post-operation (day)										
			3	4	5	6	7	8	9	10	11	12	13

Note: A. *Animal state*: (1) alive; (2) normal and active; (3) weak; (4) dead
B. *Observation*: (1) rough hair coat; (2) crusty red eyes; (3) dehydration; (4) impaired ambulation; (5) paralysis; (6) lethargy, inability to remain upright; (7) breathing: ((a) normal (b) rapid (c) very slow (d) labored); (8) vocalization; (9) inappetence; (10) bony/thin appearance; (11) emaciation; (12) muscle atrophy; (13) jaundice; (14) yellow urine; (15) diarrhea; (16) ascite; (17) wound: ((a) normal (b) bleeding (c) swelling, exudates (d) infection (e) date sutures removed (f) sutures missing); (18) transplanted heart beating: strong: h+++, medium: h++, weak: h+, stop: –

Daily Clinic Observation

Title: **Protocol**:

Animal NO	Operation date	Operation	Time points post-operation (day)										
			20	25	30	35	40	45	50	55	60	65	70

Note: A. *Animal state*: (1) alive; (2) normal and active; (3) weak; (4) dead

B. *Observation*: (1) rough hair coat; (2) crusty red eyes; (3) dehydration; (4) impaired ambulation; (5) paralysis; (6) lethargy, inability to remain upright; (7) breathing: ((a) normal (b) rapid (c) very slow (d) labored); (8) vocalization; (9) inappetence; (10) bony/thin appearance; (11) emaciation; (12) muscle atrophy; (13) jaundice; (14) yellow urine; (15) diarrhea; (16) ascite; (17) wound: ((a) normal (b) bleeding (c) swelling, exudates (d) infection (e) date sutures removed (f) sutures missing); (18) transplanted heart beating: strong: h+++, medium: h++, weak: h+, stop: –

Daily Clinic Observation

Title: **Protocol**:

Animal NO	Operation date	Operation	Time points post-operation (day)										
			75	80	85	90	95	100	105	110	115	120	

Note: A. *Animal state*: (1) alive; (2) normal and active; (3) weak; (4) dead

B. *Observation*: (1) rough hair coat; (2) crusty red eyes; (3) dehydration; (4) impaired ambulation; (5) paralysis; (6) lethargy, inability to remain upright; (7) breathing: ((a) normal (b) rapid (c) very slow (d) labored); (8) vocalization; (9) inappetence; (10) bony/thin appearance; (11) emaciation; (12) muscle atrophy; (13) jaundice; (14) yellow urine; (15) diarrhea; (16) ascite; (17) wound: ((a) normal (b) bleeding (c) swelling, exudates (d) infection (e) date sutures removed (f) sutures missing); (18) transplanted heart beating: strong: h+++, medium: h++, weak: h+, stop: –

Euthanasia Record

Date	Animal NO	Strain	Operation date	Operation	P.O.D	Anesthesia	Reason	Sample collection		Surgeon
								Blood	Tissue	

Note: A. *Type of anesthesia*: (1) *isoflurane* inhalation; (2) *pentobarbital* 50 mg (rat)-60 mg (mouse)/kg I.P

B. *Operation*: (1) *O.L.T* orthotopic liver transplantation; (2) *H.H.Tx* heterotopic heart transplantation; (3) *Skin Tx* skin transplantation; (4) *K.Tx* kidney transplantation; (5) *S.B.Tx* small bowel transplantation

C. *Reason*: (1) morbid; (2) moribund; (3) collection of samples

D *Tissue*: (1) liver; (2) heart; (3) spleen; (4) skin; (5) small intestine; (6) lymph nodes

Index

W. Gong (ed.), *Rodent Transplant Medicine*,
DOI 10.1007/978-94-017-9472-5

V

X

MIX
Papier aus verantwortungsvollen Quellen
Paper from responsible sources
FSC® C105338

If you have any concerns about our products,
you can contact us on
ProductSafety@springernature.com

In case Publisher is established outside the EU,
the EU authorized representative is:
Springer Nature Customer Service Center GmbH
Europaplatz 3, 69115 Heidelberg, Germany

Printed by Libri Plureos GmbH
in Hamburg, Germany